Reflections on Beekeeping

by W S Robson

Northern Bee Books

Reflections on Beekeeping

ISBN 978-1-904846-82-6

Published by Northern Bee Books,
Third edition 2014
Scout Bottom Farm
Mytholmroyd
Hebden Bridge HX7 5JS (UK)

Design and artwork: D&P Design and Print

Printed by Lightning Source, UK

Additional photos and editing by John Phipps, 2013.

Reflections on Beekeeping

by W S Robson

Selby Robson

In memory of my father Selby Robson

"Be good to the bees and they will be good to you"

This shows Willie retrieving a swarm from a post box down the road from the honey farm on July 1st 1993

Foreword

Willie Robson and his family have kept honeybees at Chain Bridge Honey Farm, near Berwick-on-Tweed, for three generations, and 'Reflections on Beekeeping' is the result of many years of learning, observation, and experience acquired by keeping up to 1800 colonies of indigenous bees in the North East of England and the south of Scotland. Whilst the border country has large areas of unrivalled beauty and a good range of forage plants, the wild, open moorlands where the farm has many of its apiaries are subject to severe weather conditions which demand the hardiness of both the bees and their keepers for any commercial honey enterprise to be successful.

That success has been achieved over many decades has been dependent on the long-held practices that have been established at Chain Bridge, and the family's readiness to adapt and to make changes as problems in beekeeping have arisen, especially in more recent years. The author hopes that the notes within this book will give amateur beekeepers and beginners an opportunity to learn different aspects of beekeeping and give more experienced beekeepers some ideas which they might not have otherwise have considered.

The Union Chain Bridge built in 1820. It was the first suspension bridge in the world to carry commercial traffic. Willie Robson's grand mother was born-around 1880 - in the house on the left, which was a boat house. (Picture by Stuart Cobley).

Hopefully, the book will also help academics and individuals who are interested in the countryside and its traditions to gain an insight into honeybees and beekeeping drawn from Willie's practical experience over a very long period.

As with many books on beekeeping, there is much room for thought lying behind the words which have been written, and the reader is encouraged to draw his or her own conclusions through cross-referencing and further study.

The contents of the book form a series of illustrations designed to encourage beekeepers to form a relationship with their bees. By close consideration of the material the beekeeper will realise that beekeeping is a matter of very fine judgement and that it is a great mistake to place human ideas and aspirations on to the shoulders of the honeybee.

Chainbridge Honey Farm sells a wide range of hive based products within the shop at its excellent visitors centre, though bees and queen bees are not raised for sale.

On a personal note, I remember with great pleasure my two visits to Chainbridge. On the first occasion, after crossing the lovely bridge at Horncliffe, I soon found the honey farm and was able to meet Selby and Willie, both of whom I was able to photograph alongside their doll's house beehive. Having Smith hives of my own, I was pleased to see that the farm was run entirely on this type of hive - not surprising really knowing the close association which existed between both the Smith and Robson families. Some years later, it was the visitor's centre which attracted my attention, - not only the wealth of information and displays it featured on honeybees, much of it the work of Ann Middleditch (who wrote the Glossary in this book), but also the extremely interesting archival material on the Border Country. The warmth of Northumbrian friendship and homeliness was felt on both occasions and it was good to be amongst tough northerners who were able to wrest a living from their bees in often far from ideal conditions.

John Phipps
Greece, May 2011

Top picture: A local wagon rebuilt to the last nut and bolt at Chain Bridge Honey Farm.

Middle picture: Sections of honey packed for sale to Melroses Tea and Coffee in Princes St, Edinburgh. A work of art both by the bees and the packer (Willie's mother Florence Robson).

Bottom picture: The Chain Bridge Honey Farm Visitor's centre.

The Chainbridge Honey Farm Visitor's Centre stocks honey and a wide range of hive products made on site. There are informative displays on beekeeping and Northumbrian life which will give visitors an insight into the hard life of commercial beekeepers and countrymen who take their chance against harsh moorland conditions in order to wrest a living 'from their countryside'.

Willie and Daphne Robson with a new Ibex built especially for beekeeping

Brother Adam of Buckfast Abbey, pictured here with Willie, visited Chain Bridge Honey Farm to look at the bees on the heather moors in Northumberland.

Contents

Introduction

This volume of articles has been written in response to enquiries from journalists, beekeepers and the general public about beekeeping. It is based on 50 years' practical experience of keeping bees for a living, as well as references to knowledge gained and passed down to me by my father, W S Robson Snr, who studied and taught beekeeping as a profession.

Throughout this work assumptions have been made that may not be correct, as one finds in most articles about beekeeping. Neither do we claim that at Chain Bridge Honey Farm we are always as proficient as we might be. There is always room for improvement. Nevertheless, readers must use their imagination and cross-reference to different parts of the book to pick up on recurring themes. The craft of beekeeping is vastly complicated and detailed discussions on the subject could take up many more pages. Readers must also accept that where criticisms are made I am pointing the finger at no one - I have written these notes for the general good of the craft and, by doing so, it will be seen that many of the problems that we experience today have all happened before.

Many of the conclusions I have reached will have been known about by beekeepers one hundred years ago. Unfortunately, we always ignore the hard learned lessons of the past, whatever the subject. In the last fifty years beekeeping has gone backwards in keeping with the rural economy.

It is important to be reasonably professional when attempting to look after bees. Beginners need to attach themselves to someone with a considerable degree of experience and who has been a successful beekeeper. Beekeeping is about old heads and young shoulders. In this respect it was a retrograde step to do away with many of the county beekeeping instructors who were employed by the government up until the 1970s and were a link to lessons learnt in the past as well as keeping the craft up to date.

Willie Robson, Horncliffe, June 2011.

Willie hiving a swarm.

Beekeeping Problems

Beekeeping in Northumberland has gone through difficult times, mainly as a result of bad summers (high rainfall) leading to malnutrition, and the presence in colonies of **Varroa** which can only be controlled to a degree.

Colony Collapse Disorder

Colony collapse disorder (CCD) does not occur in the North of England or Scotland but it is prevalent in America where honeybees come under great commercial pressure. The actual causes of CCD are difficult to determine but a lifetime's experience tells me that bees have lost the will to live as a community. This may be due to a number of causes. As the collapse happens during the winter, the prevailing weather is obviously a contributor or is the last straw for struggling colonies.

Continued line breeding obviously weakens bees (as it does all animals), as well as other stress factors including unsuitable transport when moving hives, the presence of harmful chemicals in the ecosystem, severe commercial exploitation to maximise honey crops, as well as pests and diseases. To me this all seems very simple. To others who deem themselves experts, CCD is still a mystery demanding a great deal of research. Great efforts are being made to find a suitable cause that can be proven beyond doubt, whereas the real cause is **bad husbandry on a grand scale where all the considerations suit the beekeeper and none suits the bees. Honeybees are not machines.**

It must also be stated that commercial beekeepers worldwide generally suffer low prices for their honey at the hands of commodity traders and beekeepers are driven to desperate measures to make a living and thus the bees are pushed to the limit as with most livestock.

Varroa

Presently, the most important thing is never to miss a treatment for Varroa. Once Varroa gets the upper hand in a colony the situation is difficult to recover. A further problem is the great number of bees being imported into the UK from Southern Europe and other places, as they are invariably susceptible to disease. Similarly, variable levels of Varroa infestation render the local bees susceptible to the existing diseases to which they were normally resistant. The problems are therefore compounded.

VARROA - brood with varroa mite which has been removed from the comb.

Nosema

Nosema is a difficult disease to control and since WWII has been a problem in the UK. Nosema has many causes, though some are difficult to determine. Bees detect apprehension and incompetence in a beekeeper, which results in bouts of stinging and bad temper. Beekeepers react by terrorising the bees with more smoke. This kind of confrontation results in increased levels of Nosema. Other causes are bad transport, particularly poorly sprung trailers, poor apiary locations (eg dampness, exposure and altitude), as well as long periods of adverse weather. Increased levels of insulation can help. Crushing bees between the supers is a certain way for colonies to get Nosema, as well as any manipulations that cause bees any stress. Whilst there are some colonies that have no resistance to Nosema, the disease is often blamed for unrelated problems in colonies that remain undetected.

My father always warned against *scumfishing* the bees, that is giving them too much smoke (*scumfish* is a corruption of the word discomfort). The word 'scumfish' was used when the crofters were forcibly removed from their houses in the Highland clearances of the 1850s and must have remained in the Scots vocabulary through to the 1950s and was then used by beekeepers.

Apiary Locations

Good apiary locations are critical to the success of a beekeeping enterprise. Bees must be sheltered from the prevailing wind, both in winter and summer. In the winter they must be able to see the sun between 12 noon and 2pm on the shortest day, without any intervening trees and bushes, to allow flight at least once a month and lift morale. Honeybees will not winter well if there are farm animals close by because of the continual disturbance, nor will they tolerate vibrations caused by heavy traffic. **Wintering bees in an area where there is stagnant air will most certainly kill them.** Consider if you will a herd of cattle that finds itself in a field that is exposed and not to their liking. They will either find the best place in the field to shelter or they will break through the fence and into a wood until they get some respite from the weather. **Beehives have to stay where their owner puts them.** It is worth noting that if northern brown bees are in direct sunlight in May and June they are liable to swarm abruptly as they are unable to control the temperature in the brood nest, especially if they are sitting on a stone slab in a walled garden.

It is most important that there is an adequate supply of pollen available for the bees at all times of year. Bees will wear themselves out looking for pollen in the spring if it is too far a distance. This is made worse by mild winters followed by cold springs. Lack of particular pollens, together with bees that are susceptible, may well be the cause of the European Foul Brood which is such a problem throughout the UK.

In order to get through the winter, bees need a honey flow in late August or September to rear healthy young bees to carry the colony through, otherwise the colony will dwindle in the spring. A good spell at the heather would help here and a bad go at the heather always spells trouble (altitude).

If a colony of honeybees stops rearing young bees during July because of atrocious weather and the risk of starvation, it will have great difficulty in getting through the following winter because the bees are of the wrong age to rebuild their numbers and make a comeback. This is a natural disaster, however there

POLLEN - sheltered sites enable bees to get out and forage for early pollen sources in cool springs.

is nothing the beekeeper can do about it, other than keep nuclei in polystyrene boxes in order to make up the colony numbers in the following spring.

Beekeepers should be aware that this is a reason for colony mortality that has nothing to do with the beekeeper or with varroa, or with pesticides. In fact the varroa are wiped out as well under such circumstances. It is often difficult for inexperienced beekeepers to pinpoint the reason for colony losses. In fact I have to think about it as well. Colony losses cause so much anguish and heartbreak among beekeepers because they are attached to the bees and very often the blame is put at the wrong door.

An easy way to get bees through the winter is to put a box of honey above the brood box. Thus the bees feel completely free from stress. If there is a queen excluder beneath the box of honey the queen will be lost through isolation in very cold weather as the bees move up into the honey store leaving the queen behind.

When bees are kept in large numbers on a site drifting can be a problem at any time of the year. Bees can drift a mile at the heather. When bees are working a honey flow in late spring during cool weather drifting can be excessive. The colonies that lose bees become demoralized and very often leave the hive altogether, apart from the queen. This is a physical problem, but the colony is often lost.

Importance Of Suitable Sites

2011 was the fifth difficult year for beekeepers in the North of England and Scotland. Bees built up wonderfully well after a very difficult winter as a result of a warm spring. Thereafter the weather deteriorated and we had rainfall almost every day and it was noticeable that the condition of the colonies deteriorated where there was insufficient natural shelter either because the field was tilted to the North or if there were no hedges. Where the hives were in satisfactory sites the bees just about held their own.

We have a few permanent sites near to the heather where the bees can be relied on to come through the winter in good condition and get a crop of honey in the worst summer. These sites are south facing, of course, and lie along the front of tall woods where there is a constant movement of air behind the hives. This means that the hives are always bone dry in winter and well ventilated in summer. Very often when the sun is at its height the hives are sheltered from direct sunlight by the tall trees. These bees are extremely reluctant to swarm. We keep about 6 hives in these sites so that there is rarely a problem with malnutrition and there is always plenty of work for the adult bees to do. This also stops the colonies from swarming.

I sometimes hear beekeepers talking of giving a super of foundation to a hive to keep them busy but it is the availability of a consistent honey flow that keeps them working and the easiest way to guarantee that is to have fewer colonies in a site. This is also a useful way to prevent them getting disease. Commercial beekeepers cannot do this, as they need to take large numbers of colonies to monocultures. This practice leads inevitably to deficiencies within the pollen spectrum which leads to brood disease unless the bees are immune to the effects of specific malnutrition which they generally are not. This is a worldwide problem, but particularly in America where commercial beekeepers take huge numbers of hives to the almond orchards for pollination services.

I remember many years ago a lady asked me if I would sell her a hive of bees. We don't sell bees as a rule but I said I would lend her one although I didn't think she would make a beekeeper. We had twenty hives on a permanent site close to the heather. It was noticeable that some colonies never did any good within this apiary and we gave her one of those thinking it wouldn't be a great loss.

We were soon to discover that there was nothing wrong with this colony as when it was placed onto its new site, on its own, it built up into a very strong colony and consistently filled boxes of honey and never swarmed despite standing in a walled garden. We had thought that the trouble was genetic but it proved the point that bees do better on their own or at least in small numbers on the same site. The initial problem had been caused by drifting and also by insufficient nutrition/forage for the number of colonies in that area and the constant lack of specific nutrients. When you get down to six hives in a site these problems are greatly reduced but not eliminated. Neither would these problems arise in big cities where the pollen and nectar sources are very varied. This is why we try to overwinter hives in or near villages where there is a great range of garden plants. Pollution, however, must be a drawback in built-up areas.

I put another hive or two in this walled garden under an apple tree against the wall. The whole garden was very damp and overgrown. Now when the leaves were off the tree and the sun was at its lowest the rays of the sun were on the front of the hive. As the year progressed and the leaves came into leaf and the boughs bent down, the hive was shaded from the sun completely. This colony never swarmed even if it was short of space. This told us that shade and ventilation were important factors in the control of swarming. This is commonly known throughout the world of beekeeping.

Another winter site, which we have close to the heather, is an abandoned cattle court and there is a tendency to leave the bees there because it is such a warm place for them. Nevertheless, every year these bees spend all summer either thinking about swarming or actually swarming or recovering from it and consequently get no honey, while hives down the road not half a mile away are all present and correct and full of honey because they have shelter from the midday sun. To add insult to injury a third of the swarmers were queenless because the queens failed to get mated because of the persistent wet weather.

As I am completing these notes during the winter of 2012/13 we have just endured an even worse season than 2011, with persistent rainfall for a period of eight months. We have a small crop of honey – about a third of our expectations - and only because the bees are spread over a huge area. But then beekeeping has always been a very risky occupation. It is not possible to sell imaginary honey! My father said, after we got no honey in 1985: *"You'll never make a beekeeper until you've seen things at their worst."* He sometimes talked about R.O.B Manley going without an income from his honey farm for 18 months during the 1930s.

The 'Carry Over' Effect

Regarding the present situation in Scotland, I have heard of similar problems over a period of ten years or so in many parts of England. I would guess that the common factor was and is the presence of varroa mites that have become difficult to treat. Added to that, a period of prolonged bad weather and large numbers of hives in one area, as well as great numbers of imports as replacements.

I imagine that similar situations exist in Europe, but with better weather. There would have been far more bees in the East of Scotland fifty years ago, but they would have been more evenly spread, being smallish colonies of thrifty bees, and the surrounding countryside would be much more supportive. The bees would be productive and a very useful part of the rural economy.

Bees, like all animals that live outside all year, suffer from 'carry over.' This means that periods of bad weather and the malnutrition associated with it will still be causing problems one year and sometimes two years later. We have just had three very difficult summers and two bad winters. Many hundreds of our colonies 'marked time' i.e. stopped developing, during May 2010, despite glorious weather. This is 'carry over.'

The polystyrene hive is the modern defence against 'carry over.' Previously we had the bee bole and the double walled hive. Less than an inch of timber provides little defence against the weather.

Double-walled hives often keep the bees at home on sunny days in the winter when they should be flying particularly after periods of heavy frost and snow. The sun's rays are unable to penetrate into the inner box. This can have disastrous consequences as they relieve themselves within the hive.

ALMOND POLLINATION - colonies which have access to only a monofloral crop will not do well - the bees need a wide spectrum of pollens to provide them with a healthy diet. In California all ground flora are cleared from the vast almond orchards.

Dysentery

I can remember that we had trouble with dysentery during the winter of 1962-3, owing to the bees being confined for months. Since then we have never had any concern about disease in our bees. Even in the winter of 1985-6 most of them came through in a weakened state. I would guess that 75% of the bees in Scotland died out in that winter as a result of Nosema. Despite a further two appalling seasons the bees in Scotland recovered very quickly to the point that by 1990 it was as if nothing had happened. Now with the continued presence of Varroa I don't have the same confidence about our own bees being able to withstand anything that the weather can come up with.

WBC HIVES - not as good as they were thought to be - the well-insulated double-walled hives do not allow the sun's rays to benefit the colony in early spring as in single-walled hives.

Foul Brood Diseases

As for the detail of the Scottish problems, European Foul Brood is about poor nutrition and genetics (Susceptibility!). This would be most easily rectified by a good summer, but the genetic problem remains. Whoever 'improved' the bees originally has caused them to lose resistance to EFB and probably Nosema and other diseases and these genes spread among the local population via the drones. EFB was virtually unknown in the North of England and Scotland all my life, although it would be endemic. I hear reports of American Foul Brood as well and I am sure that where there is EFB, AFB is not far behind.

I consider that bees can be resistant to AFB as well, given the amount of foreign honey that was thrown into tips. A cereal factory near here left open barrels of foreign honey for years on end yet none of the local bees were ever found to have AFB. This is totally natural resistance by luck more than by design. The biggest outbreaks of AFB were around honey packing plants where the owners were keeping bees of their own and feeding foreign honey to them. I have heard of this still being done, believe it or not.

My father would oversee thousands of colonies of bees in the Scottish Borders and came across AFB rarely and only when nuclei had been brought in from further south. These colonies were always treated and a three-year inspection routine followed without recurrence. My father's friend and fellow beekeeping advisor, Bob Couston, had much more trouble with AFB in Perthshire, where bees were found to be living with the disease, the colonies being resistant to it most of the time. This proved to be a difficult nut to crack and although I am talking about thirty years ago I wouldn't be surprised if the present AFB is derived from the same colonies. I wonder what happened when feral colonies contracted AFB before varroa killed most of them. They would be a persistent source of re-infection, as robbing among honeybees is extremely common. I think that honeybees are able to resist infection if they are in exceptionally good condition and genetically robust. For example, pedigree dogs are the exact opposite.

Where there is AFB, obviously something drastic has to be done about it. However, I would suggest that burning a colony with EFB is like shooting someone who has a bad cold. Treating them doesn't address the problem either. Putting them on new combs is a good idea. Doing that would also control Nosema to a degree. Throwing the brood away seems a bit drastic.

If it were me, I would make an artificial swarm so that the queen was on new combs and take the brood and young bees away to another site to rear a young queen. Better still, I would put in a queen cell from a local bee. It could be that a change of queen is all that is needed (Genetics). Thereafter, the pressure needs to be taken off. Better sites need to be found for the bees, with fewer colonies at each site. It would be preferable to have twelve hives per site, sub-divided into two lots (Nutrition again!).

This is all easier said than done, but it has to be done. Bees will generally get over these problems themselves, given time and consideration. Good sites are paramount. There was an outbreak of EFB in Scotland about twenty years ago that went largely unrecorded. This was in colonies that had become too strong for the available forage as a result of continued selection. The remedy was to introduce an unrelated queen to every colony - there were many - and the problem receded and disappeared completely.

I remember the late Tom Bradford complaining about EFB in Worcestershire in the late 1960s. Years later I spoke to the foul brood officer for that area and he told me about big colonies of bees in the orchards with three supers of honey on them that had a bad infection of EFB. The bees had committed themselves to extensive brood rearing on the back of a good honey flow and then there wasn't enough protein to build all those bodies. The ground gets dry or then the weather changes - and there you go: expanding colonies with prolific queens but inadequate nutrition. Tom said that if the bees were strong enough to fill a box of honey in May they were strong enough to eat it in June. The same goes for pollen. The problem may be the lack of specific pollens. We are talking here about the micronutrients that are necessary to build thousands of tiny bodies. Our own colonies often take a break from brood rearing during the summer. This may be a safety valve. Who knows?

Removing all the rape honey before it crystallizes in the hive causes a total reversal in the fortunes of the colony that would not occur until nearer to the end of the season. The bees are relying on some of that honey to enable them to continue expanding through a period of bad weather. Multiply that by thirty hives in a site and the situation becomes serious resulting in starvation, perhaps, or a good going bout of EFB.

We have overcome this problem by using starter strips in the honey supers which gives the bees plenty of comb to draw as well as allowing us to leave the honey on for as long as is necessary, sometime for the rest of the season as in 2007. Fortunately this resulted in a great crop of honey. The ling heather yielded heavily in the last week in August.

It is most important for bees to draw new comb in May and June. This activity boosts their morale after a long and protracted winter. Similarly, a few swarms going about do no harm either. This, together with comb building, is the highlight of their year. This is important to them and raises their morale immensely. I remember looking at bees that had been prevented from swarming for many years by a system of complex manipulations. The owner worked all day away from his bees and did not want to hear about swarms flying about that were his responsibility. This worked well enough, but the bees lost their vigour and became demoralised and eventually contracted Nosema. Any repetitive manipulation such as the nine day inspection must be done with great care and minimal disruption or they will get Nosema. Over-manipulation is the principal cause of colony mortality in apiaries used by local associations to teach beekeeping.

Brood combs more than five years old will always carry a burden of infection which may slow colony development in the spring. It is most significant that honeybees clean their hives out with their mouths and thus a varying level of infection will travel to the house bees. If the bees are going well they will overcome these problems and I have seen many a hive do very well on combs that were fifty years old. Badly soiled combs are best disposed of. A top swarm will clean and rebuild a filthy brood box in a week by taking the comb down to the midrib and starting again. The pathogens are removed completely and the house bees will be free from infection. This is the way of the bees providing themselves with new comb.

Shaken swarms are satisfactory but you wouldn't want to do that too often or the bees will contract Nosema as a result of the upset. Bees can be put onto new combs by placing a box of drawn combs above the brood box allowing the queen up. Ten days later the queen can be gently coaxed up into the top chamber by repeated puffs of smoke in the entrance over a period of five minutes. The queen can then be trapped in the top chamber by introducing a queen excluder below the top box with a full-width entrance above the excluder. This can be achieved by propping the front of the box on two thin pieces of timber in order to release the drones. This will also stop the colony from swarming. This is seamless beekeeping. A shaken swarm is not seamless beekeeping and wouldn't be necessary if there wasn't an underlying management problem.

Willie retrieving some bees and some wild comb sometime during the 1980s.

Willie's daughter Frances cutting heather comb.

Stephen Robson on the heather moors. Wooler Common, Norhumberland.

Chemicals and Honeybees

Creosote, a distillate of coal tar, was used extensively by beekeepers in days gone by. The hives and the garden shed were coated-up annually. This gave the beekeeper a feeling of well-being and didn't seem to bother the bees, but it did get into the honey. Honey takes on every taint there is: dampness, mould and so on. I looked at a very large number of shallow boxes belonging to a friend in France who runs 3,500 colonies. They were stored in an open-sided shed that was in a slightly damp hollow. He told me that there was a very fine mould in the comb that tainted his acacia honey but was not apparent in stronger tasting honeys. This situation would cost a great deal of money to rectify. This problem is very common.

Creosote has commonly been used for preserving hives - Jeff Rounce's apiary in Norfolk was regularly maintained with this chemical - and John Gleed, in Tain, even treated the hives with creosote on cold but dry days when the bees were not flying from them.

I remember going with my father to look at some bees that had glued their hives together with red bitumastic paint. Even the faces of the sections were tinged with light pink making them unsaleable. The bees had used the paint as a substitute for propolis which probably had the same effect. Honeybees keep their hives free from infection by cleaning and by the use of propolis as a disinfectant. They would need to, considering the numbers present and the temperature in the brood nest.

A beekeeper complained one day that he had put two swarms into a hive that he had taken from his garden shed and both had died immediately. My father went over there and pointed to a tin on the shelf. The chemical had migrated through the container and into the hives by molecular interchange, rendering them unusable. A similar situation occurred when some hives were stored in the bedroom of a big house where the timbers had been treated for woodworm. The bees that were put into those hives all died.

In the absence of good supplies of propolis, the bees collect paint which still hasn't dried properly and is still tacky.

Hopefully we can soon use organic acids reliably in the treatment of Varroa and so eliminate another of the great contradictions of the beekeeping world, namely that we have spent years fighting the chemical companies and now we are dependent on them to keep our bees alive. My father said: "Never put in a beehive what you wouldn't put in your own mouth". I can remember benzine being used to treat Acarine perhaps in the 1950s, but not at our premises. But then we all need medicines at sometime in our lives to keep us going, but not prophylactically. I think that is what my father meant.

If queens are being imported from Southern Europe or anywhere else and have been reared in hives treated with antibiotic prophylactically then their progeny will have lost all immunity. This poses huge difficulties for beekeepers in this country. Health certificates would be valueless in such circumstances.

Problems with Pesticides

From the mid-1950s onwards there was a period of uncontrolled use of pesticides which killed bees in great numbers for about forty years. The situation was bad in the south of England (because of the number of fruit trees), and vastly worse in the USA. A chemical called Hostathion caused beekeepers years of trouble and loss. Since then things have quietened down until we are now faced with systemic insecticides. What effect they will have only time will tell. Beekeepers and the public are rightfully suspicious, the chemical companies powerful, and the governments weak. The real effects of pesticides are felt about ten years after they are introduced.

My friend, Andrew Scobbie, refused to take bees to the raspberry pollination in Angus because of the careless use of insecticides. Both the grower and the beekeeper lost an opportunity as a result. I said to a farmer one day that some of our bees had suffered spray damage on his property. He replied that he had sprayed all his wheat fields with insecticide to control aphids. Well, the bees had been collecting honeydew in the wheat. Bees cannot live alongside insecticides without taking serious collateral damage. And what about all the other insects that are destroyed at the same time? Some of those will be beneficial. This was not an isolated incident.

I am reminded that about ten years ago a lady visited here from London who had a very responsible job in Government. She was keen to talk about the decline in beekeeping and mentioned lack of skills as well as pollution. I was aware that cadmium and lead (heavy metals) had been pollutants in city honey and I assume that this problem will now be in the past. She, however, considered that chemicals used to control garden pests in parks and gardens posed a very real threat to honeybees and other useful insects as those chemicals are cumulative and persistent (neonicotinoids). I thought at the time she was pretty near the mark.

I was concerned to read in this month's American Bee Journal about colony collapse disease highlighting certain chemicals found in beehives belonging to the writer. Some were to do with varroa control and whilst they were keeping varroa at bay, their presence – and one in particular (coumaphos) – would be affecting colony health to some degree. Also present were Chlorothalonil and Chlorpyrifos and it was assumed that the bees had carried these chemicals into the hive having visited garden plants as the bees were not in an agricultural area. A detailed discussion followed outlining the toxicity of these chemicals when they degrade and mix with other chemicals in the environment. It all made very depressing reading. Thousands of tons of these specific chemicals are used in US agriculture and horticulture annually. While I still believe honeybee problems are largely a result of incompetence on the part of the beekeeper I also accept that the presence of significant quantities of toxic chemicals in the environment are likely to be a contributory cause of Colony Collapse Disorder.

About four years ago a fellow who was involved in the marketing of agrochemicals told me that systemic insecticide was being applied to the seed of oilseed rape before it was planted and I thought this might eventually lead to trouble. Time will tell.

Agriculture, other than organic agriculture, is now at the stage where nothing grows without it being given its prescribed amount of chemical either to make it grow or to control the ever increasing levels of disease. The soil has become a medium for applying chemical, and soil fertility is a thing of the past. This situation no doubt gives us a reliable source of food but at a cost. The farmer has to get direct payments from the taxpayer to cover his costs and make a living and one of his biggest costs will be for chemicals. Then there is the cost of administrating agriculture where there are far more people employed than there are actually working in farming. There is, of course, the unknown cost to the environment. As we have seen, some of the chemicals are killing the pollinators and what else besides. Certainly, when all the costs are taken into account, food is not cheap. I can clearly remember times when agriculture was thriving, when there were no sprays and very little artificial manure and farmers relied on rotations. Food was not cheap then either but it was arguably of better quality and certainly none was wasted.

When I was a child I was aware that the river Tweed was full of fish, namely brown trout, yellow trout, grayling, roach, perch, flounders, dace, gudgeon and eels and then at some point there was none apart from the migratory fish (salmon and sea trout). You couldn't see a single living fish. I told this story to a farmer who was also an agronomist and a keen fisherman the other day. "Yes," he said, "that was the sheep dip" (organo-phosphorous). When dipping was finished for the day the left over dip was let into the nearest watercourse. He said that these fish were getting back into the rivers gradually.

I can remember there was a covey or two of partridge on every farm and many skylarks. There were corncrakes in the pasture in front of our house. They aren't there now. This is quite a deplorable state of affairs largely driven by supermarkets, the multinational chemical companies and politicians (cheap food). Such is progress.

Badgers

I was asked the other day about badgers and whether they are a threat to beehives. We have kept beehives immediately adjacent to a badger sett with no problems. No doubt most of our hives are in areas where there are badgers.

BADGERS - badgers have enough problems. Their occasional foray into an apiary is unlikely to cause much harm to the bees.

Badgers appear to have immunity to wasp stings judging by the damage they do to underground wasp nests. Badgers have presumably been raiding wasp nests for the larvae for thousands of years, whereas up until the time when beekeepers started to keep bees in hives or skeps feral bees would always be found higher up in hollow trees. Badgers would have no opportunity to rob the wild bees' nests so wouldn't build up any immunity to the stings.

At the time when young badgers are beginning to forage and explore food sources they very often drag their claws down the front of a hive and obviously provoke the bees into coming out and stinging them in the dark. The pain of the stings will cause them to leave beehives alone for good. I know there are badgers in other parts of the world that are immune to bee stings. I remember we had bees along the side of a cornfield on a farm near Jedburgh and one day we came to the hives and the corn was well flattened out in front of the hives to the extent that the farmer had noticed. We concluded that a family of badgers had visited and all the colonies had come out and set upon them and, rather than run, they had attempted to dislodge the bees by rolling about which wouldn't do them much good. Surprisingly enough the farmer had come to the same conclusion. I suppose it could have been foxes, but I think they would have left in a hurry at the first sign of trouble. Badgers are persistent foragers.

Many years ago the local hunt passed through a heather moor where we had bees and a flock of sheep ran into the hives and turned perhaps three completely over. Badgers then waited until low temperature disabled the bees. They then ate everything, completely disembowelling the hives. They only left the wire and top bars from the frames. I did not complain to the hunt because they and their friends and relations provide us with free access and an enthusiastic welcome onto their land with our bees. There will have been times when we have done stupid things on other peoples' land but they will tolerate it provided you don't make trouble for them. This understanding holds good for all the places where we work. A grouse beater made a lot of noise in front of some hives one day and got four or five stings for his effort. I phoned the landowner at night to apologise and he just laughed it off but he could easily have told me to remove the hives and not bring them back. Since then he has told us how pleased he is to have the bees on his property and he asks no rent. That is the way the countryside works.

Wasps

Wasps do not deserve their bad reputation! Wasps are omnivorous and from March to July they build up their colony numbers by killing insects and caterpillars that are found in trees and shrubberies including many well known garden pests. Wasps start their nests afresh every year and, unlike honeybees, have no reserves, so one bad month in the early part of the year will reduce the wasp population by a huge amount. It is the same with bumblebees, so a reduction in their population could just be the result of bad weather early on.

Subsequently, wasps need a sustained period of humid weather to allow greenfly to multiply in order for them to keep going and become established. It is at this time that they become beneficial insects. I have often seen a cotoneaster bush covered in honeybees, with the odd wasp running about underneath the branches trying to pick up the outline of an insect against the sunlight. So we have two similar insects working in the same bush for different reasons. The wasp is a hunter and the honeybee is a forager.

It is only during August that wasps become a nuisance because there are so many of them that starvation is inevitable. They occasionally visit beehives and swing from side to side trying to find a place where there are no guard bees on duty. Our type of bee will wrestle with them on the lighting board (at the front of the hive) until they beat a retreat. During October wasps will get into a hive after the bees have clustered and the temperature has gone too low for the bees

WASPS - a lot of good is done by wasps; however, in autumn they can be a nuisance when they have too much time on their hands with little brood to feed. When colonies are opened or at feeding time they will take every opportunity to grab a free sweet meal.

to defend. Restricted entrances will usually control this problem.

Very occasionally a complete wasp nest will decamp into a small hive of bees (nucleus) and the bees will be overwhelmed. Certainly honeybees and wasps will be very conscious of each other's existence, but the honeybee is every bit as tough as the wasp and a lot more resourceful. Nevertheless wasps are, on balance, beneficial insects.

A farmer was driving cattle along a road at a place (in the far beyond) called Wooplaw when some of the cows and calves took a detour into a young wood on the roadside. Very soon these cows and calves came back out of the wood like show jumpers and off, up the road, with their tails up and the rest of the herd joining them in a general stampede. A car coming the other way had some panel damage and the farmer got some training in middle distance running. Investigations the next day revealed a few wasps buzzing around what was left of the nest. These are the things that happen.

Further Thoughts on CCD

I have tried to pinpoint some of the problems that are making beekeeping difficult at present. This has been written to make beekeepers understand and think about what they are doing and read between the lines. Honeybees are hugely sophisticated and it therefore follows that they will need a great deal of respect and consideration. Roughshod solutions and quick fixes generally spell trouble unless the colonies are exceptionally robust. When I read in magazines about people making splits I wonder. It is all very well dividing colonies piecemeal when they are bursting with bees and the weather good. Trying it in a bad season as a method of swarm control or preparing for the heather generally spells trouble (low morale, Nosema). If they don't want to swarm then it is for a good reason. I don't like these tiny mating boxes either. Bees hate to be in small lots (Nosema). They feel vulnerable, which is presumably why the queens mate quickly.

A scientist came from America about twenty-five years ago to look at our bees with a view to importing them into the States, as they were resistant to Acarine. He was not impressed with what he saw: smallish lots cowering down in a north easterly gale. But they were all alive and went on to get some honey and make a profit (Bob Couston sent some queens over at that time). Their bees, by contrast, were suffering huge losses due to Acarine. Fortunately I have never come across Acarine in my lifetime although I hear that it may become a problem once again.

If we lose our bees then we have to buy more bees in. Therein lies the risk. Better by far to try and keep the ones that we already have and buy in as a last resort. I feel that the Americans have bred immunity out of their stocks many years ago. This ties in with general agricultural practice. If a queen breeder in trying to improve a strain of bees breeds out immunity then he will build trouble for all those keepers who buy those queens until the colonies can be satisfactorily requeened. Honeybee colonies, by contrast, strive to be immortal. They will not achieve that without immunity. Their immunity is further eroded by harsh management for economic reasons, residues of insecticides used in agriculture and then there is the Varroa mite and the winter, and there you go.

If bees are leaving the hive and their queen for good then the stimulus must be one of absolute hopelessness. In spring it is normal for old bees to die away from the hive in order to keep infection from the colony. If there is disease present they will tend to leave the hive before their time, a process known as dwindling. This is about immunity and longevity and the age of the queen and her ability to replace those that die and also about the weather conditions in the preceding months back to the previous July. As the bees that are kept in America are 'hybrids', then perhaps colony collapse disorder is an abrupt variation of the common problem of dwindling. I know that there will be a degree of empathy between adjacent colonies. When one goes, they all go. When a colony absconds due to low morale, adjacent colonies must know what is going on, presumably by sound or pheromones, and then abscond themselves. This is 'contagion' by definition and happens in America on a grand scale. It happens in our apiaries as well on a small scale and the bees are telling us that this is not a good winter site (let the bees tell you). It never happens in a satisfactory site. In Denmark they got over this problem by using polystyrene hives. Polystyrene hives exclude the wind chill. I have recently read some research which has confirmed that adjacent colonies can communicate, even in cold weather. Thus when one hive gives up the adjoining ones go too. This is most likely to happen where there is a high percentage of old bees in the colonies (low morale). A small colony with a young queen will very often survive and ignore the call to pack up because they are made up of a majority of young bees and think they have a future.

They hang on until they eventually survive the winter, completely ignoring the laws of physics. It depends on the tenacity and character of the individuals within those little lots.

Research is important, but not totally necessary. Those doing research need to have a very thorough understanding of bees and beekeeping or it will be difficult to come to any sensible conclusions. Most of the illustrations in this article are anecdotal but the logic is elementary. Research cannot help when the basic principles of good husbandry are persistently ignored.

A honeybee colony will be sentient. The bees are aware of how they have been treated and how they feel. I am referring here to the effects of transport and protracted journeys. There are always problems with livestock, but it would be better if we all tried to work with the bees instead of against them and save ourselves a lot of trouble. Honeybee husbandry is about intuition and avoiding mistakes and having made a few mistakes coming to the correct conclusions about how improvements can be made.

Steven Purves, our foreman in the new honey house. The green supers are our first venture into polysytrene. The machine pulverises oil seed rape honey in the first stage of making creamed honey without it being heated.

Nutrition

Too Many Colonies on One Site

I called on a honey farmer in the West of Scotland to see how he was getting on. He informed me that he was packing up for good and about to become a clergyman. I remarked that the area looked good for beekeeping. After further enquiries he revealed that he was keeping bees in lots of sixty because he was learning his beekeeping from a textbook written by a very learned beekeeper. I said the book was written about beekeeping in the 1930s when every second field was a clover field and the author's honey farm was 400 miles south of here where the climate is more favourable. If my friend had kept the bees in thirty lots of six, instead of three lots of sixty, he would have been rid of disease in his bees and had good honey crops. Nevertheless he had made up his mind to pack up and I couldn't persuade him otherwise.

So when you see many hives lined up together in Greece perhaps or North America it is likely that the bees in those hives will suffer from reduced ability and morale as well as increased susceptibility to disease due to specific malnutrition otherwise known as deficiency. This is a worldwide problem.

GREEK HIVES - apiaries consisting of dozens of hives are a common sight. Not only that, but they are often placed very close together. Both of these factors are not beneficial to the colonies - nor the environment.

Nectar and Pollen Sources for Honeybees

I have a friend who keeps bees in Alston, the highest market town in England. He phoned me one July day during some very poor weather. He informed me that his bees were doing rather well and I replied immediately "Ragwort." "Oh" he said, "how I hate that plant." The pollen and nectar of ragwort boosts honeybee colonies during poor weather when there is nothing else available. The inside of the hive is coated up in a dull yellow colour from the pollen and the bees thrive on it, yet this is a plant that is universally despised.

I remember the bees getting very substantial quantities of honey from the giant hogweed before it was eradicated some years ago and now we have the Himalayan balsam producing an exquisite pale honey in large quantities. If cattle broke out along the riverside the first plant they ate was the hogweed, indicating that this plant contained some trace element that they needed.

Another plant yielding honey heavily at the heather time was the common thistle. Some of the upland pastures were a shimmering blue in August and the thistle honey contaminated the heather honey. The thistle honey was next to the midrib of the comb and the heather honey, being the later honey flow, was on top.

Rosebay willowherb (also known as fireweed) was another great honey plant where woods had been cut down on all the moors. This was water white honey with ever such a delicate pink tinge, as one might expect. My point is that these are seen as nuisance plants, yet in times of extreme dearth they keep all sorts of insects alive, especially honeybees.

Large acreages of field beans have the same effect providing excellent nutrition during poor weather. Very often the brood combs would become clogged with bean pollen.The bean honey does not taste nice but it is better than ragwort honey. Thus the beans help to build up the colonies for the heather, which is so important.

During 1985 I took bees to 200 acres of field beans and such was the weather that the bees never left the hives. No beans were harvested that year either as the pods were empty. This may have been due to the lack of pollination or more likely due to the plants giving up the whole seed bearing process altogether.

There were years I remember when the bees had to be fed right into the middle of June before they got any honey, having missed the early flows due to bad weather. This underlines how tough the bees must have been that they could go through nine months of relative confinement and still fill a few supers in July. I remember being at the Highland Show and no-one had seen a queen cell by the 20th June.

The oil seed rape has changed that situation. Colonies have become much stronger and softer, particularly in the south of the UK. Beekeepers in the west of the UK will find things more difficult without oilseed rape, especially during bad weather. Oilseed rape honey finds a very good market, especially if it is free from taints and not over-heated. I remember Canadian honey being imported into this country called Honey Boy which was a mixture of oilseed rape and clover honey, prepared using the Dyce process. This product was in huge demand. With the right skills and initiative Great Britain could produce far more honey than it does for the home market.

Field beans

Oil Seed Rape

Thistle

Giant Hogweed *(Fritz Geller-Grimm)*

Himalayan Balsam

Ragwort *(Geoff Hopkinson)*

Rosebay Willowherb (Kallerna)

Characteristics of Different Breeds of Bees

When we looked through hives on the nine-day inspection we often found queen cells and we could tell the potential of the colony by the shape and distribution of the queen cells. Small and twisted cells indicated a feral or poor doer. Large proud cells indicated a colony that was worth breeding from. That had nothing to do with nutrition whereas the size of the colony had everything to do with nutrition. When I was a boy I had the opportunity to look at clover and heather sections as nearly all the amateur beekeepers produced sections. It was most interesting to note that each colony had a different pattern of sealing their sections indicating that they were all individuals with varied backgrounds. Some were of Italian origin, others supposedly Dutch or Caucasian perhaps and others would be of unknown origin. Looking in the shops today at New Zealand comb honey it is very noticeable that the sealing of the honeycomb is extremely consistent indicating that the bees that built this honeycomb are very closely related.

Queen cells

Susceptible Bees

'Susceptible' is a key word in beekeeping. It is practically impossible to keep bees that are susceptible in the UK, where every region has bad weather sooner or later. Inbreeding or line breeding can be a problem in this respect but not always. Bees have to settle down and organize themselves to survive and the beekeeper has to be careful and allow them to do this. Bees must have the will to live and do well. It is worth noting that we occasionally find very strong feral colonies that appear to have survived, although most feral colonies succumb to Varroa.

In conclusion, it must be said that Varroa and secondary infections are the principal cause of colony losses in the UK. The Varroa mite reduces the natural immunity of the bees and destroys their morale.

In the USA and other parts of the world industrial beekeeping is practiced and this creates even more problems. We are perhaps fortunate in the North of England that we have never been able to take liberties with the bees because of the harsh climate and thus, we and the bees are more able to cope with adversity. Nevertheless, beekeeping is, and will continue to be, difficult especially in temperate climates.

Defensive Behaviour

Defensive behaviour in honeybees is sometimes genetic but more often is a result of continued stand-offs with the beekeeper, whereupon the bees become habitually nasty. The bees always come off the worst in such situations, surprisingly enough. Bees will become very manageable if they are approached with some consideration and respect, and when the work is carried out fluently and with precision. It is a matter of judging their mood. And bees can detect the slightest fear.

All too often when the beekeeper goes near the hive the bees prepare for all-out war and that should not be the case. This is induced aggression and is ingrained in their corporate memory. Aggression is otherwise induced by an outcross too far. Yellow bees are much easier to handle but don't live so well in the North. Black bees will become benign if handled correctly over many years and vice versa (un-manageable).

It is worth mentioning that honeybee colonies should not be sited under overhanging trees as this tends to make them nasty: 'Bees in a wood never did any good.' Neither should they be kept on communal stands because when the beekeeper is working through a colony the guard bees in the next door hive become irritated by the vibration and they then become 'followers,' in that they follow the beekeeper everywhere and make his life a misery.

As bees bizz out wi' angry fyke.
When plundering herds asaail their byke;

Extract from Tom O' Shanter

Drifting

Drifting occurs in all apiaries. It can be influenced by the prevailing wind or large objects, such as houses or trees in the flight line. If the direction of the honey flow is to one side or behind the hives, the bees will make mistakes, and then some will go into the wrong hive just to get home, especially if it is a strong one. There is security in a big, strong hive. I am well aware that during the late summer big colonies in permanent locations can call in not only bees from neighbouring hives but also bees from a great distance away that are very often riddled with varroa. Thus the big colony dies out, even though it has managed to gather a lot of honey. Honeybee colonies obviously know the whereabouts and the status of other colonies within perhaps a two mile radius.

Drifting also occurs where colonies are left untreated for Varroa and migrate into hives in another apiary, perhaps a mile away, belonging to a different person. This can come as a nasty shock to a careful beekeeper.

Drifting can make poor queens look good and vice versa and lead the beekeeper into erroneous conclusions (Nutrition). It will cause colonies to swarm prematurely. If a colony swarms early as a result of congestion due to drifting, the colonies either side may swarm very soon after as a result of individual bees from the central colony drifting into the adjacent hives and influencing them to prepare to swarm. These bees are 'smitten by the bug'. So you might have five colonies preparing to swarm as a result of drifting.

These are the problems associated with keeping colonies in large numbers in close proximity. Beekeepers should never share apiaries. I remember before there was universal transport, a wagon was hired to take all the bees in a village to the heather. This resulted in some hives being stuffed with heather honey irrespective of condition and others getting nothing. This caused heartbreak and heated arguments among fellow beekeepers. The prevailing wind pushed all the bees to one end of the row. Colonies need to be at right angles to the honey flow, but then that is not always possible.

Willie coming home from the heather with his Austin pick-up at Ros Castle near Chillingham, Northumberland, 1964

Tom Bradford at his heather site in Wales with his Austin pick-up

Honeybee Queens

Importation of Queens

When queens are bought in from abroad, not only does one buy the useful qualities of that breed but also the drawbacks that are often overlooked. For bees to do well in this country, especially Scotland, disease resistance, longevity and thrift are of paramount importance. When bees of a foreign persuasion find themselves in the UK and come upon a prolonged period of very poor weather, they become demoralised and open to all types of disease including EFB, which I believe to be endemic. Disease knows no boundaries. Imported bees need to have ten years in which to acquire immunity and acclimatize themselves to local conditions, or die out.

When I think back to the situation in the 1950s, most of the bees were of a dark brown type, supposedly Dutch skep bees, that were resistant to the Isle of Wight disease. There was a great upsurge in beekeeping activity in the 1950s as a result of a general shortage of food and the need to supplement very low wages. Honey farms were started all over the country. Italian bees were imported to make up the numbers - much the same situation today. Unfortunately the Italian bees were difficult to overwinter without them getting dysentery which rendered the hives useless for another year until they could be sterilized. Honey farms were often closed down because of these problems. The Italian queens mated with the local bees and the first crosses were exceptionally aggressive, which put an end to village beekeeping. Previously, bees were kept in village gardens and were used to seeing people and children and dogs and so didn't sting them very often.

We can sometimes see the descendants of those Italian bees in our colonies today, as their genes would work their way right through the local population. Occasionally one comes across jet black bees which may be throwbacks to the indigenous race. There were 50 hives within the boundaries of Horncliffe, a village with 200 people. There were 200 hives belonging to mill workers surrounding the village of Cumledge (80 inhabitants). At that time agriculture was totally dependent on indigenous wild white clover and basic slag to provide fertility. I stress the word indigenous as the clover today doesn't seem to yield as profusely as the old wild white. So the bees had the sycamore, the gean, the elm, the hawthorn and the charlock in the cornfields and the thistle in the pastures as there were no sprays, then the clover. Finally, a short journey took them to the heather. Thus there would be no malnutrition in a reasonable year and the colonies would be small and resistant to EFB and other diseases in all but the very worst season; I would guess that 80 % of the success of beekeeping would be good nutrition. In those days there was adequate nutrition for a very large number of beehives although the bees were also able to withstand severe malnutrition as a result of poor weather.

Things took a big step backwards with the introduction of selective herbicides and artificial manures, but the introduction of oil seed rape in the early 1970s was of great benefit to beekeepers as it is a good early source of nectar. By that time, however, many had packed up beekeeping for good.

Italian Queen

There were great numbers of American Italian packages imported after the severe 1962-3 Winter and, whilst we did not see any of these, a friend was persuaded to take some to replenish his colonies. Many years later, we were left these bees and as we had helped to look after them in previous years, we knew what to expect. They were down to handfuls in the spring and wouldn't lift their heads in the summer and had become degenerate. The problem was Nosema and once they get that way they are difficult to get going again and never get any honey. I suppose most of them fizzle out in the winter to be replaced by tougher ones. My father said of those that pegged out in the winter: "Good riddance". Harsh winters can be a benefit to beekeepers as a means of controlling disease, although this can be hard to take at the time.

There were Italian bees and New Zealand bees imported into this locality in the 1980s. Italian drones were seen in hives miles from their apiary. The New Zealand colonies which were situated locally all gave up the ghost within three years. I read in a magazine recently of someone watching bees headed by a New Zealand queen leaving a hive for good in North Devon. I am quite sure that New Zealand bees are very good in their own country but they will not feel at home in North Devon, which is ten degrees further from the equator and on the other side of the world.

I read a letter in a magazine recently about the benefits of importing queen bees from Crete and showing a frame of brood where every cell was occupied by brood. In reality it is better to have 20% of the area of a brood comb occupied by honey and pollen (thrift) so that the bees can come through a period of dearth, the June gap perhaps, or a dry period when the plants come under stress and consequently bees come under stress too.

A full comb with pollen

The letter also talks about 'biodiversity'. Well, biodiversity in bees means to me that they are sufficiently outbred to be vigorous and disease resistant and sufficiently inbred to remain true to type, the type being the bees that belong to the locality. As mentioned previously, a large proportion of our bees in this country have been imported at some time or other so they have their own mechanisms for self improvement, as well as overcoming the effects of bad winters.

A queen from Crete mated to a local drone would just extend the weaknesses of the imported queen into the local bees as previously discussed, which in turn might lead to susceptibility, which takes a long time for a beekeeper to sort out. It is immaterial how much honey they might make under ideal conditions if they won't stand foursquare on their own feet. I admit to knowing little about genetics. If you take the Khaki Campbell duck or the Black Rock hen these breeds are far better than the sum of their parts but the parents are always the same breeds and offspring are always true to type. This might happen with bees as well but it is much more difficult to achieve.

If honeybees are brought into an area from abroad or even from another part of the country they will have great difficulty in making sensible decisions throughout the season because they are unable to recognise any of the parameters that would influence their decision making. How often have we heard of bees breeding right through a period of dearth and ending up destitute or breeding right through a heavy honey flow and using up half the honey crop to stay alive. If conditions were always good and we had the situation that existed in the first half of the 20th century, ie, whereby agriculture was sustained by millions of acres of indigenous white clover, then big hives of mainly imported bees would, by and large, do very well. It might have been possible to keep sixty hives in a site where today sixteen would be more than plenty, unless the bees had access to a reliable honey flow such as borage.

Carniolan and Buckfast Bees

In Denmark the beekeepers are bounded by a similar understanding. The queen breeders are careful to evaluate the queens they supply, as their future depends on it. The Danes have evolved a system of beekeeping quite different from ours. They need to use polystyrene hives to get their bees through the winter satisfactorily and they need to change the brood combs annually to guard against brood disease and garner some extra heather honey. Where the Danish system falls down is the amount of sugar syrup needed to sustain the bees in a bad season. It is demoralising for beekeepers to have to feed syrup in the summer, and also costly, and no amount of sugar syrup ever replaced a honey flow. The Danes do their best to minimise risk and in a good season they produce a lot more honey than we do.

A Carniolan Queen

In Northern Germany queen breeders have developed the Carniolan bee in a similar fashion, breeding out the tendency to swarm and selecting for resistance to Nosema, as do the Danes. One might hope that these highly skilled people will eventually breed bees that are resistant to Varroa. At present the Danes try to eliminate every last Varroa mite from their hives. We must do the same or lose our business and our livelihood. This situation is in marked contrast to the general thrust of this article in that as beekeepers we must try to keep bees that have achieved total, natural resistance.

If bees have to be imported into the UK it would be preferable if they came from Northern Europe. Most of the Carniolan bees that are exported are line bred by specialist queen breeders and to maintain that status queens need to be re-imported regularly to keep the strain pure, but hopefully not related, or they will go off the rails. The beekeeper then becomes dependent on the queen breeder. In America hybrids have been used for decades which tells me that inbreeding may have become universal which in turn leads to chronic weakness. This is well known in the poultry industry. I hear of problems with Nosema in Germany this winter (2010) in both Carniolan and Buckfast colonies. If the queen breeder had done his work properly this might not happen. Queen selection is very much a journey into the unknown but resistance to Nosema is vitally important. In good times honeybees will get a lot of honey whatever their background . Trying to tailor a bee that will cope with the bad times is not easy and every season and every locality is different.

We have a few Carniolan colonies crossed with our own bees. They are strong and get a lot of honey if you can keep them together. They will build queen cells without doubt and sometimes go to pieces and remain useless for the rest of the season. We, by contrast, are trying to keep bees that are stable and stay together and take advantage of a honey flow whenever it comes without a great deal of input from ourselves. Nevertheless I have great respect for the queen breeders of Northern Europe who are trying to breed immunity back into honeybees, and not before time.

I have heard of beekeepers who are keeping bees without treatment and there are colonies which are able to withstand high numbers of Varroa mites. But we, and most other beekeepers, exert commercial pressures on our bees (transport), as well as treating them and that tips the balance against their acquiring natural resistance to Varroa. It is ever increasing commercial pressure which is causing so much trouble with honeybees worldwide. I met a honey farmer from Austria (1,500 hives) who didn't treat his bees and who lost two thirds of his colonies every winter. His greatest problem was the continued presence of swarms and drones that had come from colonies that had been treated, thus negating his efforts and sacrifice. I would subscribe to the view that if there are Varroa mites in the hives at the heather or during January, getting them through the winter will be extremely difficult.

A Buckfast Queen

Addendum March 2010:

In response to a note from Clive de Bruyn about premature queen failure:

I have heard a lot of discussion about poor queens. The most obvious cause of this is persistent bad weather from June to August over a period of at least three years. The site has to be very sheltered or the queens haven't the confidence to come out to get mated. If there is a general decline in the quality of the drones and the queens, then this has to be caused by stress. This might arise from three sources: firstly, the presence of Varroa within the hive together with associated diseases; secondly, residue from the chemicals used to treat Varroa; thirdly, the presence within the ecosystem of chemicals associated with agriculture. **Queen bees stay in the hive for a long time unlike worker bees that live only for a few weeks.** The queen's exposure to chemicals is therefore greatly increased. Honeybees are a barometer of the general state of things in the countryside.

Then there is the added problem of the almost complete eradication of the feral colonies. There would perhaps be four times as many feral colonies in the countryside as there were kept ones, particularly in woodland areas. These would act as a balance within the gene pool. They would also keep as many drones as they thought fit, whereas the trend in modern beekeeping is to eliminate drone brood to a great degree. The feral colonies that do exist act as a source of re-invasion by Varroa of otherwise healthy colonies. This is a colossal reversal in the scheme of things and honeybee pollination will have declined to a huge degree. It is worth pointing out that honeybees are consistent pollinators for nine months of the year whereas other, perhaps more efficient, pollinators come and go according to the season. This is a point sometimes overlooked by the experts.

I remember beekeepers of my father's generation talking about queens going into their fifth year before being replaced. When I was in Denmark they had queens that were three years old. They change their brood combs annually so this may be a clue, in which case we need to learn to treat Varroa organically. At least that would eliminate one potential source of trouble.

I have many times heard of expensive queens being bought and successfully introduced and then thrown out from the hives a week later as if they were aliens. If, as I am told, large numbers of queens are imported annually into the UK, it follows that many colonies will have a confused identity and they will try to re-organize themselves into a state of stability as they have done for thousands of years. This means going through a period of relative instability - and stability is most important in any livestock enterprise. I therefore think that the biggest reason for premature queen failure may be genetic instability and that the bees know it. In truth, I don't really know the answer. We don't have that problem because our bees are true to type. But we will have the problems of agricultural chemicals and Varroa - both of which may be part of the answer.

I am reminded that elsewhere in the book I have written about the complexities of swarming and all the very sophisticated behaviour required to make this activity a success. I suspect that in the last few years this swarming has changed and that sometimes honeybees don't really know what they are doing. I am aware that swarms are going about with queens that are not viable and that previously this would never happen. Honeybees depend on spells of high barometric pressure (settled weather) in order to operate successfully. They cannot fly far in broken weather without using up all their energy and neither can they go through with the swarming procedure unless they are confident that the weather will be reasonable.

Unfortunately for many years now Great Britain has suffered from low pressure systems coming in from the Atlantic every 36 hours or so during high summer and this has not helped beekeeping. Queens haven't the confidence to get mated because of constant downpours and the bees are finding fault with their queens and constantly trying to replace them. If a honeybee colony suspects that there is a problem within the hive with for instance, susceptibility to disease, the bees change the queen, in the hope that a change of genes will eliminate the trouble. This is exceptionally sophisticated behaviour. Similarly one might assume that if the queen is affected by minute traces of insecticide within the hive and the bees are concerned about her ability to perform satisfactorily they will reject and replace her.

I think that these swarms that go about in high summer and never become established are swarms of hopelessness and I cannot say what will be the cause – weather, varroa or insecticide. Probably we will never know. But if bees are collecting nectar and pollen from the flowers of plants which harbour systemic insecticides it is highly likely that the queen will be affected and the bees will reject her.

A large swarm of indigenous brown bees in a maple tree, most likely headed by several virgin queens.

Further Discussions

(A) Colony Temperament

My friend Colin Weightman who first visited us here in 1949 has encouraged me to write some more about honeybee behaviour, especially their ability to make decisions. Some years ago we had a black dog and bees instinctively hate black dogs. We brought a strong nucleus home one day for no particular reason and put it in the heather bed adjacent to the pathway leading from the house. The dog passed this way very often within two yards of the hive yet never got stung. When the hive first arrived the bees would take a few days to assess their situation and not wishing to advertise their presence would keep a low profile. Thereafter a decision would be taken in the hive that the dog wasn't a threat nor were the humans that were wearing hairspray and furry collars and breathing stale alcohol and all the things that they hate. Thereafter, the discipline that everyone was to be left alone was maintained throughout the season even when it was a strong colony.

These were black bees that, had the hive been lifted and dropped, would have stung everybody within fifty yards. This is a reaction to fear. And yet the hive in the heather bed had the confidence to continue its daily business safe in the knowledge that those that came past were not a threat to them. And that message was relayed to all the adult bees that took to the wing throughout the season and that this decision was made within two or three days of the hive being put there. How clever is this? I have no doubt that the bees that were kept in cottage gardens in the old days had a similar temperament as they would allow their owners to work on them without veil or gloves. Mind you, the beekeeper would choose a day when there was a honey flow. Bees don't like to be disturbed before 10.30 am or after 4.30 pm. There were beekeepers many years ago who were carrying out the nine-day inspection after 6pm (to look for signs of swarming) and that always resulted in some severe stingings. This was very bad for the bees and the beekeeper but in those days people worked a 48 hour week so they had to do the inspections when time allowed.

Two years ago I gave an apiary demonstration to local beekeepers on a sunless day. I purposely went into a small colony very carefully and no bees attempted to sting anyone. So the bystanders' confidence grew and many of those people were novices and they crowded round. Our bees for their part aren't aggressive by nature and seeing so many people that were no longer frightened of them decided to behave themselves. What amazes me is that those bees can make this decision in an instant and communicate that information to all members of the colony. We repeated this operation on much stronger colonies where most of the bees were at home with the same result and the demonstration was a pleasure and everyone learned a lot. The bees have to be confident that they aren't under threat at the first approach and the bystanders have to be confident and relaxed as well. Failing that the bees try to instil fear in everyone present which they generally do and the demonstration becomes at worst a shambles.

During the lunch break when people were sitting around without veils some distance from the hives one gentleman got stung on the nose. I asked him if he had had a drink that day. He had not, but he had had two or three glasses of wine the previous evening. So the bees could detect that fifteen hours later and pinpoint the exact source and one only decided to do something about it. Obviously they think that the smell comes from a predator and a predator generally means annihilation. So they get in with the first punch which gets them a bad name. Years ago a contractor sprayed undergrowth on the banks of the Tweed so that fishermen could cast for salmon. Unfortunately the spray drifted through an apiary and the bees came out and stung people in the locality for a day or two. I was told that there was a chemical in the mix which caused the bees to become aggressive. There again the bees get a bad name. That spray killed a fully grown sycamore tree and a dozen poplar trees at the same time and the landowner was not happy. This sort of thing has happened to us more than once.

Whilst working with my father it was noticeable that when the bees had had enough they always picked on me and left him alone and now they leave me alone and sometimes threaten my assistant. So the bees can assess the ability and demeanour of individual beekeepers when working together even if both have decades of practical experience. I suppose this is to be expected as most animals can detect the slightest incompetence but we have to remind ourselves that these are insects.

(B) Swarming

The Bees' Need to Repopulate an Area

I am minded to talk a bit about swarming. This is a big subject. The first move towards swarming occurs in late March when the bees start to check around their near neighbours to see which ones have died during the winter. Their first concern is to bring the local population back to its previous numbers before expanding further afield. When bees swarm they leave the parent colony without a laying queen and if the young queen fails to get mated then the parent colony is lost so nothing has been achieved. In the early days of the following spring the bees from the swarm are back, checking the parent hive to see if all is well. My father thought that the swarm remembered where they had come from. He did tell me how he knew this but I have forgotten. He said that bees had a corporate memory. We sometimes lost big swarms from hives in permanent locations and this was very disappointing. My father said 'Ah well, they will be back in the spring'. We used to put an empty hive down in the exact location where we had lost the swarm and sure enough they were back at the earliest opportunity. If that hive was moved aside and another put down the second swarm came in as well such was their fervour to repatriate the old site. Honeybees at this time of the year will have made up their minds where the swarm is going up to six weeks before the event.

Absconding Behaviour

When the sun shines brightly on the front of the hives and the wind goes into the south, black bees have a habit of absconding, leaving the hive without queen cells whereupon the bees have to raise emergency queen cells. This is not really swarming as the bees have made no provision to leave, but it has a drastic effect on the hive. I remember putting bees into a stell at the heather and it became a scorching hot day. All the laying queens left on that day together with most of the bees. Where they went I never discovered. On another occasion I put several nuclei into what had been a greenhouse with no glass in it. We thought that the queens would get mated easily in such a place - which they did. On a subsequent visit all the mated queens had left as a result of too much background heat. This is known as absconding. Yellow bees are not nearly so keen to abscond and will stay in the hive until the comb collapses. I have heard of this happening in Australia. Honeycomb collapses at 105* F. In this particular case the bees were drowned in the honey (300 colonies)

Drifting and Colony Imbalance

I have already dealt with drifting being a principal cause of swarming and that drifting causes imbalance in the colony. The principal reason for imbalance is caused by the queen laying eggs in anticipation of a honey flow which doesn't occur because of the June gap or bad weather. The old foragers don't die in this situation because there is not enough work to wear them out and the young bees are surplus to requirements and become repletes and sit in the supers waiting for the order to leave. It is likely that the repletes will not be orientated to the original site so that their first orientation is to their new home thus preventing too many bees returning to their parent colony when they begin to forage. Repletes never build comb in their own hive. This is a sign that they are about to swarm. Normally young bees will secrete wax involuntarily during a honey flow but in this instance this function is suppressed until they have reached their new home. Thus they are able to build a new brood nest in a very short time as mentioned elsewhere. Their immediate concern is to become fully established before winter. This is exceptionally sophisticated behaviour. If the swarm fails to find a new home within a few days because of bad weather their mood will change from one of extreme happiness to fear and unease and then they will become extremely defensive. Many beekeepers have gone to collect their swarm and have been attacked unexpectedly because the bees are hungry and frightened as they have no home other than the branch on which they hang.

The hive that can keep its brood nest in balance throughout the season will not swarm. There is the same number of young bees coming forward as are dying. One might think that such a colony would not get much honey but they often get much more honey than one might expect. No honey is used up raising bees for whom there is no work and it might take 100 lb every season just to maintain the colony. In 1985 the colonies couldn't maintain themselves. When these stable colonies sense high atmospheric pressure they prevent the queen from laying any more eggs and push all the bees that are able to fly out on foraging duties. They deliberately contract the brood nest (see reference to 'contraction' in glossary) thus increasing their work force by perhaps 50%. High atmospheric pressure means easy flying so the aged foragers live longer which makes the workforce even stronger. The bees work on the balance of probability and if there is high pressure there is likely to be a heavy honey flow. In 2010 our bees got a big crop of honey which we guessed was from extra- floral nectaries possibly from field beans that had gone past flowering during July (not honeydew). Certainly there were no flowers to be seen and yet a fair proportion of this crop was comb honey. The downside of all this once the honey flow is over is that the colonies become extremely depleted in numbers and it takes three weeks for them to get back to full strength. The queen resumes her normal egg-laying pattern and the colony prepares for winter unless it is taken to the heather.

The Influence of Weather on Swarming

When bees are engaged in swarming they will be aware of the risks involved. If the weather is bad they might change their mind and pull down the queen cells knowing starvation awaits any swarms that leave the parent colony. If the weather is variable they might go ahead out of frustration (imbalance). If the weather is better than average they will go ahead knowing that the swarm has a very good chance of surviving and if the weather is settled and there is a honey flow they might banish all thoughts of swarming and work really hard knowing that they might not get a chance like this again for some considerable time (years). Decisions are being made every day in the hive and there must be occasions when there is dissension within the ranks. I have seen a swarm leave one day and half of them come back the next day because they have lost interest in swarming particularly if there is plenty of honey to get.

Habitual Swarmers

So to recap, swarming, assuming the bees have plenty space, is about the urge to reoccupy vacant hives as a result of winter losses, secondly about the inability to control the temperature in the brood nest and thirdly about an overrun in the brood nest and imbalance. This is normal swarming as the surplus young bees are the basis of the swarm. Another cause of swarming is habit, as with the Dutch skep bees and the Carniolan. Skep bees were constrained in the skeps and had to fill the empty skeps from the season before which was the only way the beekeeper could get any honey, thus swarming became habitual. This was crude beekeeping. On the North York moors a schoolmaster called Austin Hyde collected skep bees in September and sent them on the train as packages to Tom Bradford in Worcestershire who hived them and fed them up for winter. In the spring the hives with the skep bees were put into orchards in Kent or locally to provide pollination and then sold on to beekeepers who had lost hives the previous winter. Thus, the skep bees lived to fly another day. This activity was known as enterprise and went on into the early 60's.

In the early years we increased our colony numbers by making artificial swarms. While this enabled us to make use of some excellent queen cells it also meant that we were breeding bees that were more inclined to swarm (habitual) which we knew was the exact opposite of what we should have been doing.

If the bees are not pleased with their queen or she is old they might supersede her or they might take the opportunity to swarm with the young queen when she emerges. And then there are bees that become hell bent on swarming, that is throwing out numerous casts that travel around the countryside until they fall foul of the autumn weather or they get into a hive. Interestingly if these small lots survive the winter they will adopt the same lifestyle as the year before - ie building cells and swarming at the earliest opportunity without attempting to get any honey. This behaviour might be a throwback to the times when little lots roamed the subtropical forests getting some honey and moving on. In normal circumstances a colony headed by a young queen will not swarm in its first season unless the colony has been heavily influenced by other factors mentioned previously, but if the young queen has headed a cast or second swarm she will go again the next spring. This skittish behaviour is carried over in their corporate memory into the following season.

If the bees during the season throw the drones and drone brood out as a result of poor weather it means that keeping the drones would bring them close to starvation and that they won't be swarming during that season, and that they are happy with their queen and won't be replacing her. If the bees retain their drones after the first of September it means that they are not happy with their queen or they are queenless.

Usurper Bees

About 40 years ago we were looking through a hive and found the queen balled (suffocated) by bees that did not quite match the shade of those in the hive. There was a small swarm (cast) hanging nearby and sure enough it was them. My father knew what was going on. When we returned, the cast had gone into the hive during the honey flow and then gone away with a big lot of bees because they were hell bent on continuing their nomadic lifestyle. These were usurpers and normally they would stay in the big hive and work once they have got in because they are generally desperate for somewhere to go. Doubtless the virgin queen would be killed on the way in. Sometimes they go in and every last bee from the swarm is killed in the hive and thrown out. The host colony changes its mind. This looks to the beekeeper like spray damage.

Every colony of honeybees has an individual character; some are robust defenders and others are more accommodating. I have heard of honeybee colonies that welcome wasps, the most notable headed by imported queens (suggesting lack of instinct?). It all depends on the intensity of the honey flow and the nature of the colony being invaded. Usurpers can only go to work during heavy honey flows. To go into a hive, kill the queen, bring in their own queen and then, having influenced the workers, depart with a big swarm is very clever and malevolent behaviour. At this time (late 1960s) honeybees were operating at the peak of their faculties. Now they are struggling to survive. I had forgotten about this until I read an article in the American Bee Journal about it. I am afraid there will be many other things about beekeeping which I have forgotten due to commercial pressures - which is why I am writing all this down.

An Unusual Method of Dealing with Swarms

We were very friendly with people called Stobie who had a small honey farm at Lilliesleaf. They produced clover sections only and had about 40 hives in the midst of a stock farming locality. The seasons were late and swarming time coincided with the main honey flow. These people were shepherds and the father watched for swarms during the height of the day and boxed them. In the evening the swarms were distributed right along the line of bee hives irrespective of where they had come from. All the hives were ripening honey and readily welcomed a few more thousand bees on the landing board. This exercise was repeated daily until no more swarms emerged. As it would be well into July the bees wouldn't be too keen to leave a heavy honey flow and would readily accept this style of management. It was quite normal for a hive to produce 100 clover sections. The hives that had laying queens continued to do well and the hives that had no laying queens were full of bees and had no brood to feed so they did better. This would go on until artificial nitrogen was used on the surrounding fields. I am not recommending this style of beekeeping, just recording that it did happen and it was very successful.

Further Note:

Despite my notes above on the complexities of swarming and all the very sophisticated behaviour required to make this activity a success. I suspect that in the last few years this swarming has changed and that sometimes honeybees don't really know what they are doing. I am aware that swarms are going about with queens that are not viable and that previously this would never happen. Honeybees depend on spells of high barometric pressure (settled weather) in order to operate successfully. They cannot fly far in broken weather without using up all their energy and neither can they go through with the swarming procedure unless they are confident that the weather will be reasonable.

(C) Artificial Swarms

To make an artificial swarm the old queen must be put into a brood box on the original floor on the original site without any brood. If foundation is used in the new box it is essential to place the two outside combs from the original brood box into the middle of the foundation making sure there is no brood on them. The queen can then start laying straight away on her own combs.

The queen excluder must be put back with the supers on top. Above the supers there needs to be an inner cover with a swivelling entrance to the rear and with the feed hole completely closed and above this the brood chamber with the developing queen cells. When the queen cells are fully developed and sealed the top brood chamber can be lifted onto a floor two or three yards away. The flying bees will return and they can be run into the bottom box, i.e. the artificial swarm, by moving it forward slightly on its floor so that an entrance appears at the rear and the bees are admitted into the new colony. The rear entrance can be closed after a few days.

A natural swarm has no brood therefore it should follow that an artificial swarm has no brood so as to be as close to the normal swarm behaviour as possible.

I will elaborate further. If a frame of brood is put into the new brood chamber with the queen on it, as has been suggested in some text books, the bees are liable to continue with their original intention to swarm or they may simply delay the initiative and start again in a week or two. They may decide that having gained some space to work in they will ignore the fact that they have a frame of brood and continue to work satisfactorily. There are many factors influencing bees at this time of year and every colony is different but if a colony has made up its mind to swarm then the old queen must be divorced completely from her brood.

The textbooks would advocate putting a frame of brood in with the queen on it so that the beekeeper didn't have to handle the queen. Picking up the queen and putting her in the new box of frames is never easy because when the queen realises that she is the object of the beekeepers attention she starts to run. If the swarming date is drawing near she will have thinned right down so that she can fly the distance required and this enables her to run faster. Sometimes it is easier to remove two frames from the new chamber and shake her and her attendants into the box making sure not to bump anything and then replacing the two frames, queen excluder and supers quickly so that the inside of the hive is dark. It is a misconception to think that putting a frame of brood in with her would help to keep her there – it would be more likely to make her leave. As an aside it is interesting to note that very occasionally a queen will faint or feign death when picked up after a bit of a chase. Big black beetles that sometimes inhabit the space above the crown board regularly feign death when picked up so they aren't stupid. No earwigs or other scavengers live there if these beetles are in residence

This may be a good time to clip one wing of the queen if it hasn't been already done. We never marked the queens as my father considered this practice as an unnecessary insult. On reflection I would probably agree. So we spent a lot of times looking for queens. We used to find them by looking down the face of the next comb to the one we had just lifted out.

A farm manager who had got the beekeeping bug phoned my father one day to tell him that he had found the virgin queen in the hive and had clipped her wing. Father then asked him how he expected the queen to get mated. History does not record what was said after that!

The Repletes

It is essential that the supers go with the new box as they hold the repletes, and it is the repletes that carry the swarm forward through the days between the flying bees that go with the swarm dying, and the first of the new generation hatching and reaching maturity. The repletes sit in the supers of the parent hive full of honey and not doing any work thus extending their lifespan. Once they arrive in their new home or their artificial home they will draw a deep box in a few days. They already have the honey to do it in their honey sacs. If the supers are removed and put on top of the original brood some distance away the artificial swarm will find it impossible to maintain the necessary momentum to develop their new brood nest.

It is important initially to put the original brood on top of the board above the supers as the heat from the activity below helps the nurse bees to make a good job of the cells, as well as look after the very considerable amount of unsealed brood. Willie Smith and others kept very strong colonies of bees by building brood chambers up this way so that the heat from the chamber below helped to rear the brood above. Readers may have noticed that in bad seasons bees will follow heat up through the supers and put the honey in a chimney formation, always following the highest temperatures. This enable them to access the honey easily during the winter. They are very clever.

When the brood is moved away to a new site the flying bees that might have contributed to a second swarm will drift down the back of the hive and move in with the artificial swarm. This will relieve the beekeeper from having to reduce the original brood box down to one queen cell. It is good to leave bees alone as much as possible. If a small cast comes out then so be it.

Experience

My father and I made thousands of artificial swarms over the years and perhaps 85% would be successful. If we were doing it today then we might be lucky to achieve this rate of success because bees are much keener to do away with their queens. If the artificial swarm fails and the queen has gone it is a simple matter to make a three-frame nucleus on the old site leaving one queen cell and no open brood and move the rest of the brood away two or three yards. It is always better to have two units going in case one queen doesn't get mated. It is important for beekeepers to learn to recognize supersedure cells (one only or very occasionally two) in which case no further action need be taken.

It is no use making artificial swarms with old queens (feathery wings). A quick look at the brood pattern will determine the quality of the queen.

Reasons for Failure

Beekeepers ask why, when they have successfully made an artificial swarm, the bees start again and build more cells. This is most unusual. Perhaps the bees wish to do away with their queen. I can only guess at other reasons. A beekeeper who collects swarms in the Newcastle area in an official capacity told me recently that of fourteen swarms collected seven had queens that were not viable. Why would bees fly about looking for a new home when their queen was not viable? This is irrational behaviour and bees are always rational, or they were in the past.

If they thought that their queen would not see them through the next winter they didn't leave their hive. They waited until a young queen was born and went with her. It is very worrying to note that our bees no longer seem to know what they are doing. So making satisfactory artificial swarms today is going to be difficult, simply because the bees no longer seem to want their queens. I find this quite alarming and I suspect a chemical influence, but then I don't know.

Other reasons for failure

If bees are uncomfortable in their site due to strong sunlight and the lack of ventilation they would be liable to swarm again. This would be particularly true of black bees. Otherwise there may be a degree of instability within the colony caused by bees being imported and then out-crossing with local bees or others that have been imported. When a colony of bees finds itself out of step with the climate and district in which they live as well as possessing other abnormalities such as the lack of resistance to disease they embark on a protracted period of self determination in order to bring themselves back into line. This is fundamental to their very existence. In this case I think that bees are always learning and memorizing the necessary alterations to their behaviour in order to ensure the maximum chance of survival. They will also be changing their queen at every opportunity to hasten the journey to normality. Thus we have artificial swarms building cells for a second time in one season. In one word – instability.

I suppose if bees were swarming in April they might go again in June but I can't help thinking that strange behaviour is the result of some chemical influence or bees being moved over long distances as well as variable levels of inbreeding. A clever queen breeder can produce queens that are very suitable for honey production in the UK but they can never match up to the honeybees' ability to tailor their genetics to their environment so that they are still there in a hundred or a thousand years. Honeybees will also adapt to a beekeeper's style of management over a period of thirty years or so and if they are treated with consideration will improve accordingly. If the beekeeper has a heavy hand either they will die in the first bad winter or go into a state of permanent non co-operation until another beekeeper takes them on. If the bees have been in the care of a good beekeeper they will not readily adapt to another keeper.

(D) Heather Honey Production

Heather Sites

In my experience the flowering date of the ling heather can vary between 27th July and the 21st August depending on the weather during the growing season. Drought will bring the date forward into July and the season is sometimes all over by the glorious 12th August, which is the traditional date for taking the bees to the high ground as well as the start of the grouse shooting. I find that if we can get the bees into perfect shelter at the heather there is a much greater chance of getting some honey. This means getting them down into places where the air is always still. It takes time to find these sites and if the weather has previously been poor then it is the only way to get a crop. The bees must feel inclined to expand whereas the slightest exposure will cause them to contract and fill the brood boxes with honey, with the added risk of fewer young bees for the following spring.

Heather site at Primrose Cottage, Debdon near Rothbury, Northumberland, on an estate belonging to the Armstrong family who were industrialists on Tyneside in the late 19th cenury.

Some of you, in Scotland anyway, will remember Steele and Brodie and perhaps Nathaniel Grieve in Edinburgh. Both of these firms manufactured heather floors where the bees ran in at ground level (this was important) and ran up a sloping board within the hive and up through a narrow slit at the back into a second level immediately below the brood frames. The normal entrance was blocked off completely so that no cold air could enter the hive. I remember looking at hives that were stocked with Caucasian bees and they had built a curtain of propolis right along the mouth of the hive for the same purpose (propolis – pro polis meaning 'before the city'). In 2011 we found that bees that were taken to the heather and put into perfect shelter did well while others not so well sited got absolutely nothing. It is all about colony morale.

Poor heather sites are also the cause of a great deal of queenlessness in the following spring. Many colonies like to change their queens at the heather before the winter (supersedure). If the site is exposed the queens will not come out to get mated. Extreme shelter and still air are of paramount importance at the heather.

When choosing a site it is important to imagine where a flash flood might occur after a downpour. I once lost twenty hives that way. I found some of them and others ended up miles away. My father told me that during the 1947 flood there were hives of bees banging up against the bridges over the Whiteadder water in Berwickshire. Not long after that the bridges went too.

Managing Bees on the Heather

Beekeepers Stephen Robson and Joth Hankinson moving beehives to the heather at Debdon Moor near Rothbury, Northumberland. using one of the two custom built Ibexs we possess. We have a 4 wheel and a 6 wheel, both with differential locks, to ensure the vehicles are not stuck in the mud.

When bees are taken to the heather many foragers are lost as a result of flying out into an unfamiliar locality. When a colony loses bees like that under normal circumstances they will expand the brood nest to make good the numbers. Occasionally at the heather they will do the opposite and prepare for winter. They find themselves in what they consider to be a very poor location and with winter approaching they batten down the hatches, putting any honey they might get into the brood chamber. If those hives are left too long on the heather and then go into a poor winter site they won't be alive in the spring. This is a big problem in Scotland. If they get into an outstanding winter site they will generally rally and get going again. Producing heather honey depends on the bees being in good condition when the heather is at its peak (dusting) and the weather is good. This is a tall order and explains why heather honey is expensive. Dusting is when the heather pollen rises if the plant is disturbed.

Drifting and Robbing

As mentioned previously, drifting can be a problem at the heather. Initially the bees go into the tallest hive or the one at the end of the row. Thereafter if the heather flow is directly in front of the hive and not round the corner, certain colonies are able to influence bees from their neighbouring colony into leaving a lot of their honey with them. Whether they are able to exert this influence when they are flying in the corridors approaching the hive or on the lighting board I am not sure. I suspect that some colonies are more assertive and know that a bit of robbery without violence could be useful in the run up to winter. I am also assuming the communication would be chemical. I noticed that this had been discussed by my father and Willie Smith in 1963 so they knew that it took place but weren't sure of the method employed by the robber bees. They also mentioned that ling heather would be at its optimum for honey production at three years old. Currently we are taking bees to heather that is fifteen years old because the owner has been forbidden to burn it.

Conservation and Management of Heather Moors

In my lifetime about half of the ling heather has been destroyed in the North by reclamation and forestry. Some of the land that was reclaimed was so poor that the farmer went out of business trying to repay the capital cost. The land would have been better left as heather. When there was a great surge in forestry in the 1970's, ling heather was regarded as a useless plant and trees were planted in huge swathes obliterating great areas of moorland habitat. Trees should have been planted with regard to the contours and varieties planted which, when mature, might have provided some amenity and value as timber. And then trees need to be looked after and thinned, and that means hard work.

The government, represented by Natural England, are now involved in managing what is left of our heather moors and as they have little idea of what they are doing I fear for the future of this magnificent plant. Management of heather moors and the sheep that belong there will have been known about for 1,000 years and its practice is cast in stone. Yet there are people in universities where theories abound that know better than our ancestors, using money that currently doesn't exist and with the backing of our friends in Brussels. Ling heather needs to be burned regularly for it to thrive, the period between burnings being the life of a sheep - about six years. Currently the sheep are being removed and the shepherds made redundant under Government orders. This is a scandal as communities and skills are destroyed and irreparable damage will have been done.

Willie's son Stephen tending bees at the heather at Thrunton Crag near Whittingham.

Willie's 1966 VW pick-ups and, in the background, the first honey house that he and his father worked from.

Willie with the next model up of VW pick-up (1980s) and a big load of cut comb!

(E) Beekeeping Education

I believe it was a retrograde step in the 1970s for The Ministry of Agriculture to dispense with the services of the county beekeeping instructors. Beekeeping instructors gained their knowledge from studying the relative success or failure of those they taught, from novices to professionals. Once knowledge is lost the threads are rarely, if ever, retrieved. Thus we now have a situation of uncertainty when people are trying to pick up skills, worsened when things go wrong.

It may be difficult to avert trouble during the present problems affecting beekeeping, but it would be easier if there was a body of men and women with decades of practical experience to instil some confidence in the profession. This is not to decry the efforts of the people at the beekeeping unit in York and Auchincruive, whose contribution I appreciate. It is very important to look back, but they look forward and have useful information for the present and the future.

For forty years after WWII, Scotland had a beekeeping education facility which would have compared favourably with any in the world. Since then the countryside has been nationalised and the emphasis placed on administration together with suitable legislation. Such is progress! The only bonus for beekeepers from this situation is that there is now a large acreage of oilseed rape grown, which was a political crop, but is now valued by farmers as a break crop.

Willie and his wife Daphne meeting the Queen in Berwick Town Hall in 2002

(F) Bees Do Nothing Invariably

The literal meaning of this saying is that beekeepers may carry out manipulations "by the book" and the bees for reasons which may or may not be apparent will do something totally different as a reminder to the beekeeper that he is not totally in charge and never will be.

If we use our imagination and take our thoughts a step further it could mean that as honeybees have colonized three continents unaided and come through ice ages and endured numerous other setbacks they are unlikely to be impressed by demands put upon them by humans. Hard won independence is the very key to their existence and beekeepers must learn to recognize this or trouble surely follows as we have clearly seen.

My father often said "Be good to the bees and they will be good to you". This simple statement might be attributed to the late W.W.Smith of Innerleithen and there will have been times when I and plenty others will not have been so good to our bees as we might have been. However the real meaning of this saying is: yes, don't leave them without any food and moreover leave them alone as much as possible and they will reward the beekeeper in due course with fewer stings and some honey.

This discussion reminds me of a saying "The farmer's foot is the best dung". This meant that the farmer should walk around his fields often and see what was happening both with his crops and livestock.

Just imagine a farmer working in the depression of the 1930's or maybe 100 years before that. He would have no help from the banks or the government, no fertiliser or chemical sprays, little or no mechanisation and a rent to pay. All he had was his ability to farm and he would derive that from acute observation, intuition and experience. This would also hold true of beekeeping and particularly commercial beekeeping even today, whereas farming has become much less difficult because of modern techniques and support.

Old Time Beekeepers

I remember years ago when there were cottage hives for section production. Bees don't like sections, especially in the spring and with a zinc excluder in place. The bees often filled five frames with brood and then left with a top swarm. If there was somebody at home to throw the swarm back into the hive the queen could be killed as she ran back in, and the stock reduced to one queen cell at the required date. Generally the bees were not fed in autumn so that those that had done well and filled their supers died out and those that gave nothing, but filled the brood chamber, survived. This resulted in a very poor quality bee, quite useless in fact (In the 1950's beekeepers were given a sugar ration specifically for feeding to the bees - some bees never saw it!).

However, there were beekeepers who selected the best hives every year and reared a few queens and united them back to the poorer ones and so began a system of selective breeding. The late Willie Smith of Innerleithen is an obvious example of a progressive beekeeper. He was Scotland's first commercial beekeeper and the inventor of the Smith Hive. The late George Hood had his bees, and it shows. The late Alec Cossar of Kelso gave them extensive bottom ventilation, an empty deep in fact and he always had a few queens ready for the heather. He was a gamekeeper, so he had the time to do it, which was critical.

The late Rob Brown of Pallinsburn had big WBC hives, all in a row, outside the gardens where he was in charge. His bees would work sections from May onwards. He eliminated the swarming tendency by removing queens that persistently built cells and breeding off the ones that didn't. He also eliminated queens that bred bees that didn't seal the sections to show standard, believe it or not! My point is that these beekeepers could produce some quite excellent colonies of bees from the ones they already had. When nuclei are united to strong colonies they become habitually and genetically stronger, just as the little section hives were habitual swarmers, same week every year and so on. Black bees, or brown bees as I like to call them, accurately reflect the ability of their keeper.

The queens that progressive beekeepers raised would often be mated with a close relation or sometimes to a feral drone or one out of a cottage hive and that prevented things going wrong as a result of inbreeding. Drones are known to migrate considerable distances in order to prevent inbreeding within the indigenous population. They presumably are allowed into colonies on their journey for bed and breakfast before moving on. Occasionally colonies do become inbred and lose their vigour, especially ones that always supersede. Inbreeding would be a precursor to any disease. This is well known throughout agriculture.

After WW2 there were large numbers of black bees imported into Scotland from France by Steele and Brodie of Wormit, Fife. This race of bee still exists, I believe. It might well have been the same bee as the British Black. Steele and Brodie had obviously done their homework, as these were good bees. Their progeny will exist throughout Scotland today. Steele and Brodie had a vested interest in providing good bees to local beekeepers, as they supplied them with all their beekeeping requirements. Break that trust and the whole lot goes west. This is in marked contrast to the trade in nuclei of indeterminate provenance, where the seller has no interest in the subsequent fortunes of the beekeeper. This makes life very difficult for beginners. This went on when I was a boy and it still does today.

W W Smith of Innerleithen

L – R Willie Smith, Selby Robson, Mrs W Smith, Willie Robson, Mrs S Robson, 1960.

Willie Smith was Scotland's first commercial beekeeper and the designer of the Smith hive. My father did not think that he was given enough credit for what he had achieved. Willie was a despatch rider in the first war in France. Thereafter he became a chauffeur for Mr Ballantyne, a mill owner in Innerleithen.

Mr Ballantyne encouraged Willie to keep bees during the daytime when he wasn't needed as a driver. I would imagine he would have cottage hives and as he wished to be a progressive beekeeper he designed a new hive which was made according to British Standard dimensions but to the American pattern. The hive was very cheap to make being four pieces of wood only. He went for top bee space as this prevented the box above coming into contact with the lugs of the frames below allowing the box to be tipped up and inspections made for queen cells; thus the frames in the bottom box were not dislodged by the upward force and no bees were crushed. This would be a problem with metal ends which were used at that time but would diminish with the introduction of Hoffman frames. The only drawbacks of the Smith hive are the poor handholds, and a deep box weighs 70 lb when filled. His hives were made of close-grained white pine and painted with white lead paint as in the American tradition. We still have one at home.

In the spring Willie put on a brood chamber of frames from which the heather honey had been scraped. Peebles, where Willie had his bees, is a late area and this provided a huge boost for the bees. Readers may know of the American tradition to provide pollen patties in the spring. Willie's method amounted to the same thing. Thereafter, the queen had three boxes to work in and I suspect the bottom box would be largely brood-free thus providing ventilation and some considerable control of swarming. Normally the queens were replaced by the supersedure impulse but if they did build cells he would find the queen and make an artificial swarm.

It wouldn't be easy to find the queen in three boxes although their wings would be clipped and the queen marked. At a certain date the queen would be put down into the bottom two boxes. This is known as contraction.

Sometimes the queens built cells in July out of frustration due to unsettled weather. This provided him with huge problems as he needed to keep the bees together. It meant going through every hive and shaking the bees off the combs and removing the cells until the bees lost interest in swarming. By this time the hives were very powerful and Willie only had a net veil and bare hands. This is unimaginably difficult work. On one occasion R O B Manley and A W Gale of Marlborough visited Willie during the summer and my father got them off the train. When they arrived at the apiary they were attacked by bees and Willie could not be found. Needless to say he had retreated into the bushes. In days gone by when protective gear was rudimentary in the extreme it was very common for beekeepers to retreat into the undergrowth to get rid of followers. On that day the bees were not out of hand, just extremely peppery. This caused a good deal of amusement. I remember being at a demonstration at Kelso where the bees belonged to Alec Cossar, another exceptional beekeeper. On that day Willie Smith was speaking and a bee flew into his mouth and stung him on the tongue. He barely interrupted his speech. People were more than impressed.

Willie's usual mode of attire was a heavy tweed suit with a waistcoat, a collar and tie and a light net veil on the brim of his hat. He was a very big strong man and worked 120 hives single handed. I cannot think that he moved many about especially as he worked double brood chambers with very often two brood chambers on top as supers. In Peebleshire at that time the bell heather came nearly to the valley bottom with the ling on the tops. Permanent sites therefore made a great deal of sense. Since then Peebleshire has suffered from afforrestation on a massive scale. It would be difficult to run that honey farm in that area nowadays.

Willie Smith's skyscraper hives. This photograph bears lasting testimony to a Scottish beekeeper of quite exceptional ability. (Photograph by the late George Hood.)

When the supers of heather honey were brought home he scraped the honey from the frames into muslin bags and pressed it. His target was 5 cwt. a day. Thereafter, the jars of heather honey were put into warm water baths and sold as clear heather honey with air bells. The grocer in Innerleithen dressed the complete shop window with his honey. This was often photographed.

His honey house was on the banks of the Leithen Water and, as I remember, was hexagonal in shape. My father persuaded him to give a talk at a conference in Northumberland and he had to sit through a lecture given by somebody who did not know what he was talking about. This made him extremely annoyed and as he had a heart condition my father got him pushed away round the corner before he could confront the chap. Willie was extremely intolerant of people who didn't know their job.

My father was brought up in a beekeeping family. His grandfather had sixty skeps but he was trained as an agricultural botanist. He joined the Edinburgh College to teach beekeeping in 1949 and as they had only kept cottage hives he needed to learn about progressive management. Thus he and Willie Smith became great friends. Willie's bees would be selected from the local strain and yet R O B Manley wrote to my father saying he had never encountered such powerful colonies. This spoke volumes for Willie Smith's ability and the plentiful flora in that locality at that time. I remember them talking about queens going into their fifth year before they were replaced.

George Hood got some of his bees when Willie retired and the rest were taken over by George Lunn who was an apiarist at the college. George Hood retained an apiary in Peebleshire and the bees are still there.

Willie Smith read books written by American beekeepers in the early part of the 20th century in order to gain knowledge about commercial beekeeping. No doubt he would have cottage hives and would select what he considered to be good colonies and rear queens, make nuclei and unite them back to the parent colonies. Thus his particular strain of the local bee would become more prolific and would fill more than one brood chamber with brood until he got them to the stage where they would fill three chambers with brood. They were not two-queen colonies. The object was to have massive colonies ready for the heather. The great honey farms in the south of England would keep mainly imported bees in large Dadant hives yet their colonies were not as strong as Smith's and his bees were of a local indigenous strain entirely acclimatised to Peebleshire, a late and difficult area for beekeeping.

Willie Smith kept 150 hives of bees reducing to 120 hives nearing the end of his forty- year commercial beekeeping career. In the photograph Willie Smith is on the right. The best hives were in six British Standard deep boxes. This was not unusual as I witnessed on several occasions.

George Hood

George Hood was a great friend of our family over a period of fifty years. I remember as a youngster being impressed by the quality of his sycamore honey. East Lothian has always been a grand place to keep bees and there are parts away from the coast that are heavily wooded and contain many sycamore trees. Large quantities of this honey could be obtained in a good season, mainly I think from extrafloral nectaries, because the flowers do not last long. When the oil seed rape started in the mid 1970s the bees were inclined to ignore the sycamore and go to an easier nectar source.

About one year in ten we used to get a box of honey per hive off the hawthorn during the last week in May. This was dark, aromatic honey of the most exceptional quality. When the Northumberland coastal strip was turned over to cereal production these hawthorn hedges, which were six metres high were cut down and a useful nectar source lost.

At one stage George looked after 350 hives on his own and these were big, four-storey hives like those that belonged to Willie Smith. They were kept in lots of twelve and he stuck to the nine-day inspection to prevent swarming. They stayed in their winter sites for the summer honeyflows and then all were taken to the heather during July. He took some a bit earlier to catch the high ground clover. The hills were very close by. I can't imagine how he got through the work, but he did.

He phoned me about once a week, generally early in the morning, asking why I wasn't out of bed (he started at 6am) and we had far-ranging discussions, mainly about beekeeping, which I greatly enjoyed. I miss those phone calls now because he could certainly brighten up the day. Maybe it was just as well he wasn't here in 2012.

Some years ago he was telling me on a weekly basis that he could no longer cope with the huge amount of work and offered me 200 hives of bees. Well, at that time I hadn't got the back up that I have now and I thought that if he couldn't cope how was I going to manage with all the bees that I already had? So I turned the opportunity down. This was a decision that I have come to regret. He really wanted me to look after those bees and, given that our colony numbers have been falling lately, we could have done with them.

I am sure that honey farmers the world over when they become overwhelmed by the amount of work at the height of the season become irrational and dispirited. But then there is no option but to get on and see the season through and make the best of it. Beekeepers suffer from low morale as well as the bees!

George, having been brought up in a mining area south of Edinburgh, was well able to stand his ground. Once he was delivering honey into a warehouse belonging to Jenners department store in Princes St, Edinburgh. Now the store man at this place was always in a bad mood in the morning but in the afternoon, after he had had a few pints of beer at dinnertime, he was all sweetness and light. Well, George would deliver in the morning and loaded his honey onto a barrow and into the store, whereupon the store man pushed it back into the street. George pushed it back in and it came out, in, out, in, out all accompanied by coarse and appropriate language. Meanwhile George's mother who had come along for the ride had become upset by this altercation and had to be comforted by a bystander who got into the cab beside her. Unfortunately he hadn't had a wash for a while which upset her even more. In this case George won the day and the store man backed off.

Like most beefarmers George was a master of the 'insecure load'. He bought some hives from Andrew Scobbie in Kirkcaldy, Fife, and some of the roofs blew of when he was crossing the Forth Road Bridge. The traffic following would have had an exciting time.

George had his own accordion dance band for many years. This was his winter occupation. He loved accordion music. He was friendly with Sir Jimmy Shand who was leader of the most famous Scottish country dance band. He was famous because his music lifted people's spirits like none other and he kept perfect time which was so important for dancing. On one occasion Jimmy Shand was playing for some dancers who were performing in front of the Queen and the Duke of Edinburgh. No doubt the man in charge was in a bit of a fluster because he told Jimmy Shand the wrong number of bars of music to coincide with the end of the dance. After this shambles was finished the Duke of Edinburgh stepped up to Jimmy Shand and said "Well, you got that wrong." There wouldn't be a reply.

I went once with my parents to a dance where he was playing and it was difficult to move about never mind dance there were so many people there. I was told that when he went to Australia it was standing room only wherever he went. This has not got much to do with beekeeping except that George Hood was a friend of Jimmy Shand, and he talked a lot about accordion music and I was happy to listen.

Alec Cossar

Alec Cossar was a salmon fisherman (ghillie) at Kelso where the River Teviot joins the River Tweed. Readers may have heard of the internationally famous Junction Pool. He worked there, as well as other beats close by, and he would meet many people from all walks of life. He was able to talk to them at their own level, which was so important, and he was very well liked and respected.

He generally wore, whether at work or otherwise, a light green tweed suit and plus fours, a waistcoat, collar and tie and a flat tweed cap and heavy brogues. This was his uniform so that anyone would know his occupation.

Alec Cossar was a quite outstanding beekeeper keeping ten colonies of the local strain of bees in National hives on double brood chambers. He had a third brood chamber filled with ten frames of foundation placed underneath to provide ventilation and clustering space. The bees did not work on this

Sir Jimmy Shand on the left and George Hood playing the accordion.
(Courtesy of Stuart Hood)

foundation as they worked upwards, the foundation was sacrificed, and the box would be re-used every year until the foundation disintegrated. This stopped the bees from swarming and he never looked at his bees for queen cells. He kept a note of the age of the queens and re-queened the colonies regularly by uniting nuclei to them so that they were always brimful of bees and in tiptop condition for the heather. I remember his bees being as strong as those belonging to Willie Smith.

He was at a great disadvantage to Smith, as Smith kept his bees in sheltered permanent locations close to the heather so that the switch from flower honey production to heather honey was seamless and no adult bees were lost because they couldn't find their way home. Alec Cossar, by contrast, had to take his hives to the heather some distance away in the Lammermuirs and he had to put his bees where the estate vehicle could go and the site would never be ideal. Many adult bees would be lost through flying out into unfamiliar territory, and a further drawback would be that the heather might not be yielding nectar when he took the bees and he might be a fortnight out in his guess as to when this might occur. Smith's bees however would know to within half an hour of the onset of the heather honey flow and in the meantime would stay at home and keep their condition.

A further problem for Cossar's bees was that they would have to endure very low night temperatures which would be made much worse by a light north westerly wind that blows at night in these valleys. His hives had ¾ inch entrances and he used carpet felt as a closure. He found out quite by accident that if he left the felt in and only eased a corner out the bees did much better. He only produced cut comb as he could sell it without needing any machinery, but then again cut comb isn't easy to produce in a poor season. Nevertheless in a good season he got a lot of heather honey.

Alec Cossar used to talk about 'wrinkles' . This was about making difficult and laborious jobs much easier without upsetting the bees. I have looked in the dictionary as I write this and a wrinkle is a 'valuable hint' (Scots dialect).

He had a great sense of humour. When a farmer phoned up and told him that some young stock had got into his apiary and knocked over a few hives he commented: "They must be devils for punishment."

During an apiary meeting one of the onlookers started complaining and cursing because a bee or bees had got into his clothing and were making their presence felt. After the meeting was over Alec remonstrated with the chap for swearing when the minister was present. "I didn't notice the minister because he was wearing a veil" came the reply. At the next meeting Cossar stepped forward and said in a loud voice "Now afore we start is the minister here?" Everyone fell about laughing - except perhaps the culprit. The minister was the Rev. John Hall, the father of Anne Middleditch who has helped me write all my articles.

Willie Kirkup

For many years a fellow called Willie Kirkup worked for us. He kept us going with equipment everyday during the summer without fail so that we were ready for the off in the morning. He kept fifty hives of his own, despite having a serious heart condition, and during the early 1960s he bought some Italian queens to replenish his stocks. Needless to say the colonies headed by Italian queens died out during the winter whereupon he bought some more Italian queens and so on. When Italian bees have to endure a winter in north Northumberland they shiver to keep warm, their metabolism working more than it should do at that time of year. This, coupled with the fact they are living on heather honey, causes them to need to go to the toilet and because the temperatures are generally low they relieve themselves in the hive. Our black bees however cluster tightly and can go for longer periods without needing to fly.

My father spoke to Willie Kirkup one day and told him in a stern manner not to waste his money on any more Italian bees and look after his own bees. From that day on his beekeeping flourished by keeping six colonies per site within a five mile radius so that they always had adequate nutrition. Eventually his profits became an embarrassment for such a small enterprise because the hives cost him nothing to keep. He never forgot to praise my father for his sound advice. I know full well that commercial beekeepers cannot reasonably work with such small numbers of colonies and that Italian bees are fine for Italy and similar climates.

Willie Kirkup was a rabbit catcher to trade. He paid a rent to the farmer and often that money paid the rent of the farm. The farmers had low rents because most landowners cared greatly about the community and provided jobs and free houses for life and security for many hundreds of people. Some estates locally still adhere to these principles.

The rabbits were caught in snares in the early morning and taken to the nearest railway station and put into the guard's van and taken to game dealers in the industrial Midlands the same day. This was lucrative business until the Inland Revenue started to look into it. One morning he thought he might miss the train and as he had a perishable cargo thought he might chance crossing the A1 (the main road between London and Edinburgh) without stopping to save a few minutes. As luck would have it he struck a wagon loaded with potatoes and his van was wrecked and the contents (rabbits, cartridges etc) were spread all over the road. He was taken to hospital with head injuries. He discharged himself from there in the afternoon to go and look for his dog which had disappeared into the countryside. At the subsequent court hearing he overheard the magistrates discussing a fine of £4. But the chairwoman, one Lady Tankerville, disagreed, saying that it should be £6 as 'this could have been a serious accident'. A weekly wage at that time was £4,10 shillings or 10 new pence per hour.

He always used a Morris van for his beekeeping. I asked him once why he didn't have a pickup and he said he didn't want anyone to know what he was carrying. This is a typical countryman's response. However vans aren't much good for transporting bees. A farmer told me that he had seen Willie driving along the road with a curtain of bees hanging inside the windscreen and he and his assistant both had veils on. On another occasion the sump of his van struck a rock whilst crossing some moorland and the hives burst open in the back and the bees attacked him from the rear. He and his assistant then lay face down in the heather for a while, until things quietened down. His assistant refused to go near the van again, so he had to unload the hives himself and then take the fellow to hospital. Later that day he got a message from the assistant's wife saying that her husband would not be helping him again. Willie Kirkup was a bit reckless. He would get this way during the war, but he did produce a lot of honey and was a great help to us when we needed him most.

Tom Bradford

Tom Bradford had a honey farm of about 450 colonies at a place called Castle Morton near Malvern in Worcestershire. He kept bees in three counties, namely Herefordshire, Worcestershire and Gloucestershire. He kept his hives in orchards, some of which belonged to him.

Tom was very well respected by everyone; he was a good beekeeper and an astute businessman. He did a great deal of work for the British Beekeepers Association, often helping them when others would have been looking after their bees. He liked to socialize with other beekeepers, talking to them in an abrupt, authoritarian manner. Nearly everyone liked him. He was a character. He really did know it all. I found him an inspiration.

He wore a charcoal grey suit with collar and tie if he was on official business, and the same but without a collar if he was working on bees. George Hood went to visit him to further his knowledge having been taught beekeeping to a very high standard by Willie Smith. They went out to put supers onto the hives and at one apiary there were several swarms hanging in the trees. Having put the supers on Tom made ready to leave. "What about these swarms?" says George. "We ain't got no flippin' time to bother with them!" says Tom. "If we don't get these supers on there will be a lot more like that." Not a conventional decision, but the right one all the same. George was amused.

In a good year he got seventeen tons of honey and his honey extractor stood high up in the honey house with the motor mounted in the loft above and a flat belt coming through the ceiling. Thereafter the extracted honey ran into an OAC (Ontario Agricultural College) strainer consisting of concentric filters, the largest on the inside and the finest on the outside, and then into gold lacquered honey tins. All his honey was sold to local shops under his own label and it was mainly dark tree honey.

If visitors arrived when he was extracting honey he pressed a large, serrated knife into their hand and a pail and told them to un-cap honeycombs while they talked. This they did. The wax cappings were put through a spin dryer (hydro extractor) to remove the last of the honey and they came out as dry as sawdust. He used coconut matting as a filter because it was springy and didn't choke up solid. We still use these machines today. They are dangerous if they get out of balance, which sometimes happens. It is a matter of listening to the note. If the note changes, look out for big trouble!

Tom took bees to the heather in Wales in an Austin A40 pick-up, fifteen hives at a time. There were six hives in the bottom and a further nine on the top with the outer floorboards resting on the edge of the bodywork. This made a compact load, but was a bit risky. On one occasion his helper, a youth, was late in the morning and Tom was really annoyed. After they got loaded, to make up the time, they took a short cut through some narrow lanes. There they came upon men digging in the roadside, no doubt some of our friends from the Emerald Isle. As the pick-up bounced over the heaps of soil a hive came loose and fell off. "We ain't stopping," says Tom, fearing a showdown with the Irishmen. That would certainly put a stop to their digging for a while and some local beekeeper would get a hive of bees for nothing.

Tom Bradford always gave a very detailed lecture and a lady in the audience, wondering where he got all his information from, asked if he read many books. He replied that he did not but if he got books he generally gave them to the bees to read! This was a cheeky reply to a cheeky question.

Tom's wife, Pat, used to rear calves for Kidderminster market. That was her little business within the bigger operation. What a privilege it was to know these most resourceful, independent people.

Combs built by bees inside a skep held by Tom Bradford.

Willie (left) and Tom Bradford at his honey farm at Castle Morton, 1961.
(Apologies for the quality)

Encouraging Beginners

by Selby Robson

Many years ago I was at a large Agricultural and Horticultural show in Edinburgh in the company of an official of the Dept. of Agriculture for Scotland. As we approached the Beekeeping section we saw a large banner which read 'Beekeeping for Pleasure and Profit'. My friend said 'Pleasure maybe but not profit – no, there is no profit in beekeeping'. I strongly disagreed with him because I knew many gardeners, gamekeepers and other rural workers who, in those days of very low wages, relied on sales of honey from their hives to provide some of the luxuries they would not otherwise have been able to afford.

In those far off days, before the introduction of chemical sprays on farm land, it was comparatively easy to get good crops of honey without much effort as there was usually a continuous succession of nectar-bearing plants during summer such as charlock in the cornfields followed by white clover in the meadows and finally heather on the moors. Although modern farming practice has destroyed many of the plants on which bees depend, it is still possible with careful management to get reasonable crops of honey in most seasons.

The role of the honeybee in nature is as a pollinator, and some surplus honey and beeswax are little more than welcome by-products of the great work that honeybees do in the pollination of fruit and seed crops and the tremendous contribution they make to the food supplies of mankind throughout the world.

All our fruits and many of our seed crops depend on insect pollination and where these crops are grown on a large scale there are not enough wild pollinators to do the job. A perfectly-developed apple must contain ten seeds, which necessitates perfect pollination, so we have thousands of hives of bees moved into Kent from neighbouring counties for fruit pollination and the growers gladly pay more than £20 per stock for their services. In our own area farmers clamour for bees to be put in their rape and bean fields for pollination, and in the USA some bee farmers depend entirely on fees for pollination work, moving their hives to successive crops throughout the season.

In recent development where early strawberries are grown in polythene tunnels, placing a hive of bees in the tunnel has been found to increase the quality and quantity of the crop by 100%.

So anyone with a hive of bees in their garden should expect to get top quality fruit. They will also greatly benefit their fruit growing neighbours within a radius of a mile or so.

Before deciding to take up beekeeping as a hobby one should give the matter careful consideration. To purchase a stock of bees without knowing anything about them or their management will most likely lead to disappointment and disaster. One of the best ways to begin beekeeping is to join a local beekeepers association. Go to the meetings and talk to the members; one will find them very friendly and helpful to beginners. Ask one of them if they will allow you to help them with their beekeeping – to be their 'Gofor' for a season. Having seen something of the management of the bees and that one doesn't react too badly to a sting or two, one can proceed to purchase a nucleus stock the following season and help it to build up to full strength.

One will find that one has acquired a very interesting hobby with some surprises and much pleasure and, hopefully, some honey and profit.

Tributes to William Selby Robson by John Gleed:

Scottish correspondent. from *The Beekeepers Quarterly*, Spring 1990, No 21.

Mr William Selby Robson was 84 years of age when he died on the 1st of February this year. This was sad news, but at the same time I reflected with pleasure on the many occasions I was in his company for one reason or another. I first met him in the immediate post-war years when he was the Beekeeping Advisor for the Borders and had his office in Newtown St. Boswells. At that time he had a white pick-up for transport and even now I remember that the letters of the registration number were OPX. I cannot recall the figures, but OPX was enough, and many a time over the following years I would scan the vehicles in Kelso Square and when I saw the OPX I knew there was every chance I'd fall in with him and be able to talk bees. Looking back on this, I must have been an awful pest and surely Selby would have been forgiven had he said to himself when he saw me bearing down on him, "Oh no, not this bloke again!" I remember sitting with him in that same pick-up in Kelso Square one winter evening with snow deep on the ground as he questioned me in connection with a beekeeping proficiency certificate for the Scottish Beekeepers' Association. One year he held a series of six lectures in the village school at Eccles, a village seven miles from my home. This was in January and February and every Monday evening I cycled that round trip of 14 miles to sit

(Selby left) and Willie Robson standing in front of Willie's first Unimog.

in the school desk seat as he unfolded the mysteries of the bees to us. Such was his ability to instruct, hold interest and instil enthusiasm that I could not wait for these Mondays to come round. The beekeepers in the Borders owed him a big debt of gratitude. My lasting recollection of Selby Robson was his beekeeping attire. An old Burberry raincoat, a cut hat and a simple black net veil. The outfit may have been basic in the extreme, especially by today's standards, but the beekeeping ability was all there. It is more than thirty years since I last saw him but recently, just a matter of months indeed, I saw a photograph of him and even now I could still see the same man 1 knew so well. His son Willie, is a prominent member in the beekeeping world today, and to Mrs Robson, Willie and the family, we do indeed extend our sympathies. For me, I have many pleasant memories of the man who led me into this craft and I am glad to be given this opportunity now of expressing my sincere appreciation of him.

As bees flee home wi' lades o' treasure.
The minutes wing'd that way wi' pleasure

Extract from Tom O'Shanter
Robert Burns

Glossary

by Ann Middleditch

Absconding – when all or part of a colony leaves the hive..

Acarine disease – a disease affecting the tracheae (breathing tubes) of the honeybee. The disease is caused by a tiny mite – *Acarapis woodi* – which infests the tracheae where the complete life cycle of the mite takes place. Crawling and dying bees may indicate acarine but could also point to a viral disease.

American Foul Brood – AFB – this is a serious bacterial disease of the brood of the honeybee. i.e. it affects the larval (grub) stage of the bee. It is a notifiable disease and must be reported to the authorities if suspected. There is no treatment and affected colonies must be destroyed by burning along with the frames. The outer boxes must be scorched to kill off any bacteria. This is carried out by the bee disease officer.

Apiary – a place where bees are kept.

Artificial swarm - swarming is the natural way for a colony to reproduce with half the colony leaving the hive to set up home in a new site. To prevent losing a swarm a beekeeper may divide a colony which has already started swarm preparations but has not yet left the hive. This is called an artificial swarm.

Balling of the queen – if for any reason the bees don't take to their queen they may kill her by suffocation by forming a tight ball around her. The ball is about the size of a golf ball. This can happen when a new queen is introduced to a colony without the beekeeper taking proper precautions such as using a slow-release queen cage.

Basic slag – a bi-product of the steel industry, rich in lime and used as a slow-release fertiliser on the fields to promote the growth of wild white clover. The clover fixes atmospheric nitrogen in order to grow grass to feed animals that produce manure to grow crops. This is the type of agriculture that existed in the early part of the 20th century that supported so many hives of bees.

Bee bole – a recess to give protection to the straw skeps. Some were quite simple being built into stone or brick walls of buildings. Others were very elaborate structures built of stone or wood with spaces for several skeps.

Beeswax - honeybees secrete beeswax from specialised wax glands in the abdomen. They construct their entire nest from honeycomb made from the beeswax. The honeycombs hangs vertically. Honeycomb has a central midrib and hexagonal cells on both sides. The cells are used for food storage and for rearing brood. A wild bees nest might have several parallel combs hanging vertically in an enclosed space. In a beehive the bees build the honeycomb within the wooden frames which the beekeeper puts in the hive.

Benzene – at one time the treatment for acarine disease was to fumigate the colony with a product containing benzene. This is no longer done and acarine is not regarded as seriously as it once was. At present there is no authorised treatment for acarine.

Bottom ventilation – in very hot weather a beekeeper may prop up the brood box to allow more air to circulate to prevent overheating. An empty brood box placed between the hive and the floor serves the same purpose.

Brood – the development stages of the honeybee – egg, larva, pupa.

Carniolan bees – the race of bee originally from Austria, Yugoslavia, and other countries of the Danube Valley as far as the Black Sea.

Cast – a second or after swarm which emerges after the main swarm has left. It will be headed by a virgin queen.

Caste – in the study of insects this is the term used to describe individuals of the same sex but with anatomical and behavioural differences. In the honeybee colony there are two castes, the queen and the worker. They are both female but their anatomy and behaviour are very different. There is also a male bee called the drone.

Charlock and thistles – both common plants of agricultural land. They supported honeybees and other insects but as they were 'weeds' they had to be eradicated.

Colony Collapse Disorder – CCD – this term is used to describe the unexpected sudden death or collapse of a colony. There seems to be no single specific cause but may be a combination of factors such as the use of pesticides, succession of bad summers and winters, varroa infestation, stress. The problem is worse in the US where commercial beekeeping methods put the bees under great stress.

Contraction - contracting reducing the area in which the queen can lay thus restricting the brood nest. This takes bees away from house duties and pushes them onto foraging. Bees that belong to the locality will often do this themselves but imported bees will not. This is important if honey flows are few and far between as in the North.

Creosote – use of – many years ago beekeepers painted their hives with creosote to preserve the wood. Because it is thought to be carcinogenic it is no longer available. The powerful smell could affect the taste of the honey that will pick up taint easily.

Defensive behaviour –if a colony of bees feels it is under attack it will defend its nest, stores and brood by stinging the intruder. This defensive behaviour is a means of survival.

Double-walled hive – design of hive which has inner boxes to hold the frames and house the bees and outer boxes (or 'lifts') to protect them. They are often of telescopic design typical of the 'cottage style' hive. Gives good insulation but expensive to buy or construct and too heavy for migratory beekeeping.

Drawn Comb – this means that bees have 'drawn out' or built the wax honeycomb cells onto the sheets of foundation that the beekeeper has placed in frames in the hive.

Drifting – bees may go into the wrong hive on returning from foraging. Causes could be prevailing wind blowing the bees off course, hive layout – the end hive in a long line may collect extra bees.

Dutch skep bees – the use of skeps in Holland continued for much longer than in this country. The Dutch skep was taller and more conical in shape than the UK skep. The Dutch beekeepers bred bees suitable for this kind of beekeeping.

Dyce process – a method of processing honey where it is heated to melt the crystals then cooled rapidly and stirred. The result is a smooth creamy spreadable honey. E J Dyce an American professor devised the process which is now used throughout the world

Dysentery – a symptom of a disease such as Nosema or a nutritional disorder due to fermented stores. In summer bees defecate away from the hive so the problem goes unnoticed. In winter if the bees are restricted in flying they may defecate on the combs making them unusable.

Emergency queen cells – if a colony suddenly loses its queen the bees will build an emergency queen cell using a young larva in a worker cell. The bees extend the worker cell vertically to accommodate the queen larva. Usually built in the centre of the brood nest rather than on the edges of the comb as in swarm cells.

Empty deep – by 'deep' a beekeeper means a deep box to hold the brood combs as opposed to a 'shallow' for the stored honey. An empty deep placed under the brood box helps with ventilation in very hot weather. If a colony overheats it may swarm or abscond completely and combs have been known to melt in extreme cases.

European Foul Brood – EFB – a bacterial disease of the brood. It rarely kills a colony but will weaken it. It is a notifiable disease and must be reported if suspected. It can be treated with antibiotics by the bee officer but if badly affected will be destroyed.

Extra-floral nectaries – some plants will produce nectar from nectaries not associated with flowers. These can be found on leaf bases, leaves or stems. The nectar is the same as that found in flowers and bees will collect it and produce normal honey from it. Field beans, laurel and cherry are examples.

Feral colonies – colonies of bees living in the wild.

Feral drone – a drone (male bee) from a feral colony.

Frame – beekeepers use rectangular frames consisting of a top bar, side and bottom bars. The bees build their comb within the frames so the beekeeper can control where comb is built. Also he can remove individual frames easily.

Foundation – thin sheets of beeswax imprinted with hexagons to give the bees a guide where to build comb. The beekeeper can thus control where the bees build comb.

Habitual swarmer – a colony which swarms frequently. The beekeeper would try to eliminate this trait by requeening from a non-swarmy stock.

Honeycomb – see Beeswax

Honeydew - this is the sweet sticky substance excreted by aphids and other tiny insects which feed on plant sap. The honeydew is collected by the honeybees and stored as honey. It is referred to as 'honeydew honey' and can vary in colour from golden to almost black.

Indigenous race – the local native strain of bee in a particular area. The honeybee has a wide distribution throughout Africa and Europe. All the bees from different regions belong to the same species – Apis mellifera . However distinct races did develop in different regions as they evolved to cope with local conditions. These differences are either anatomical or behavioural. For example the Caucasian bee of eastern Europe has a longer tongue to reach the nectar in the local flora – red clover – and the Carniolan bee can tolerate severe winters and builds up quickly in spring. Northern races have hairier bodies to keep them warm. There were no indigenous bees in America, Australia or New Zealand. They were taken there by beekeepers.

Isle of Wight disease – IOW – in the early part of the 20th C disease decimated thousands of colonies of bees in Britain. The problem started on the Isle of Wight in 1904 hence the name. In the 1920's the acarine mite was discovered and immediately (and erroneously) was blamed for the IOW disease. For many years acarine disease was referred to as the IOW disease.

June gap – a period of dearth of nectar between the late spring nectar flow and later flowering plants. Important for beekeepers to be aware of this as colonies can starve if the early honey crop has been removed.

Line breeding – this means breeding from selected colonies in the apiary. These colonies display the characteristics which the beekeeper wants, for example, good honey production, little swarming, good temper. Queens are reared from these colonies and then used to requeen less useful ones. The whole stock in the apiary can be improved greatly. However, continual line breeding can eventually lead to in-breeding which would weaken stocks. To prevent this, periodically unrelated stocks should be added to the apiary.

Marking and clipping of queens - a beekeeper may mark the queen with a spot of suitable paint on her thorax just behind the head. This enables him to find the queen more easily among the thousands of workers. Some beekeepers also clip the queen's wing on one side. This prevents her flying off with a swarm. She will fall on the grass and may be lost. The swarming bees will return to the hive once they realise the queen is no longer with them so the beekeeper doesn't lose half the colony and has time to take steps to prevent the colony swarming again.

Mating boxes – tiny hives holding just a handful of bees and a queen cell/ queen, sited away from the apiary until the queen is mated and laying. Used by some queen breeders as not many bees are needed for each small hive.

Midrib - see Beeswax

Nosema – a disease of the adult bee involving the digestive system, caused by a spore. In severe cases it can so weaken the colony it is useless for honey production. One sign is if a colony does not expand normally in spring.

Nucleus – nuclei – a small colony usually 3 to 5 frames of bees. Suitable for beginners to start with. Commercial suppliers usually sell nuclei rather than full size colonies.

Packages – package bees – colonies of bees with laying queen transported in boxes without frames for the beekeeper to transfer into his empty hives. Beekeepers in the US have been buying in package bees from *Australia to restock their hives after suffering severe losses in the last few years.*

Pollination – the transfer of pollen from the male part of the flower to the female part of the same flower or one of the same species. Pollination is the precursor of fertilisation of the seed or fruit. Self-pollination is when the pollen is transferred to the same flower or one on the same plant. Cross-pollination means the pollen is transferred to a flower on a different plant but the same species. Self-pollination can lead to inbreeding so cross-pollination is preferable.

Pollination is carried out by insects, wind, water or animals e.g. birds, bats. Wind and insects are the most common pollinators with honeybees being one of the most important of the insect pollinators.

Pollinator – honeybees are very efficient pollinators and pollination is the most important part of their activities. They actively collect pollen for their diet and their hairy bodies easily transfers the pollen from flower to flower. They are 'flower constant; which means they tend to stick to one type of plant while collecting nectar and pollen. Also as they live through the winter they are ready in big numbers to pollinate the early spring flowers. A large proportion of our food relies on bees for pollination.

Polystyrene hives –the excellent insulation properties of the material keeps the bees warm in cold climates.

Propolis – propolis, which has antibiotic and anti-fungicidal properties, is a natural product made by bees from the resin of trees and plants. Bees collect the resin from sticky buds and tree bark and mix it with beeswax and other substances. Honeybees use the propolis to seal any small holes or cracks in their nest and to varnish all the surfaces within the hive and because of this it is said the inside of a bees nest is one of the most hygienic places known to nature. Propolis can be used in ointments and other pharmaceutical products because of its healing properties. Propolis is a mixture of resins, beeswax, essential oils, pollen and organic matter plus minerals, vitamins and other minor components.. The resins make up about 50% of propolis with beeswax about 30%. Propolis is rich in trace elements such as iron, copper, manganese and zinc.

Prophylactically - this means medication is administered to a colony even if it hasn't been established that it is needed. The thinking is that it will keep disease at bay but in actual fact the result is the colony loses its ability to fight disease.

Queen breeder – a beekeeper who rears queen as a commercial enterprise, selling the queens, when mated and laying, to other beekeepers.

Queen cells – a developing queen is raised by the workers in a special cell – large, acorn-shaped and hanging vertically - which gives the developing queen space to grow.

Queen excluder– a flat screen with slots or gaps placed between the brood box and honey super . The workers can pass through the gaps but not the (larger) queen thus restricting her to the brood box. This prevents her laying in the honeycomb. Excluders can be simple sheets of zinc or plastic with slots or a better design with a wooden frame and wired grid.

Repletes – Some bees will store honey in their honey stomach as a food store in preparation for swarming. These bees are referred to as repletes. Other insects such as ants have repletes in their colony.

Secondary infections – these follow the main infection because the bees are already under stress and so succumb to other diseases e.g. a colony with varroa will then show symptoms of other conditions such as paralysis virus.

Sections – small square open-ended boxes, 11 x 11 x 4.7 cm that the bees fill with honeycomb. First introduced in 1857 they were a popular way for beekeepers to produce honeycomb. Round sections known as cobana were introduced some years ago but the production of sections has more or less died out apart from a few enthusiasts.

Section hive – beekeepers specialising in section production would aim to breed bees that fill perfect sections. Any colony not up to standard would be requeened with a queen reared from a good colony.

Section production – to get the bees to fill the small boxes (which they generally dislike) they have to be crowded together to force them to use the sections. This can lead to swarming which the beekeeper wants to avoid as he loses half his workforce.

Shook swarms – a method of moving a colony of bees into a new box of clean combs. Each frame in turn is removed from the old box and the bees shaken into the new one. It would be carried out if disease such as nosema was present in order to try to eliminate the infection. The manipulation is stressful for the bees.

Skep – a dome-shaped hive made from twisted straw or interwoven wicker. Skeps were used by beekeepers for centuries until the introduction of the wooden hive with moveable frames.

Splits – some beekeepers split or divide a colony in order to increase the stock. This can stress the bees if they are not in swarming mood.

Starter strips – narrow strips of foundation fixed to the top bar of the shallow frames to give the bees a guide as to where to build their honeycomb. Ideal for 'cut comb' (where the honey and wax is eaten) or the comb can be crushed and the honey and wax separated.

Stell – a stone-built shelter for sheep, usually a circular wall with a gap for the sheep to enter. Sometimes a stell is roofed in.

Super – the honey box placed above the brood box (hence 'super') where the bees store their honey. Bees always store honey above their brood, even in the wild, which makes it easier for the beekeeper to remove surplus honey without disturbing the bees too much.

Swarm – bees reproduce by 'swarming' i.e. half the colony leaves the hive with the queen and sets up home elsewhere and the half left at home raise a new queen.

Varroa – a parasitic mite that lives on honeybees. Its natural host is the Asian honeybee that can tolerate the mite. The mite spread through Europe and reached the UK in 1992. It is now widespread in the UK and beekeepers have to treat their colonies to keep mite numbers within acceptable levels as our honeybee cannot control mite numbers by itself. High mite numbers causes stress and secondary infections will cause the collapse of the colony.

WBC – double-walled hive designed by William Broughton Carr. His initials give the hive its name!

Yellow bees – black bees – different races of bees have different characteristics such as colour, length of tongue, number of hairs on body, size etc. These characteristics develop as bees evolve to suit their local conditions. The Italian bee has for example yellow bands on its body and the Northern European honeybee is dark.

Zinc excluder – a queen excluder made from a sheet of zinc with slots to allow the workers through but not the queen. Wired excluders are more popular with beekeepers.

Postscript

I am told by beekeeper friend that his bees in the Tyne Valley have become partially resistant to varroa mite and that feral colonies are surviving in his area and I am sure that he is correct in his assetion that honeybees are learning to clear mites from their hive like their close cousins *Apis cerana*.

Closing Words

These notes are derived from observations made by my father and his circle of beekeeping friends some of whom were very knowledgeable particularly Willie Smith and observations made during the 25 years that my father and I worked together and observations made since. Despite the reponsibilities of running a business there is always time to see what you see. Honeybee colonies are corporate bodies that behave like individuals and they are constantly making decisions and do nothing invariably. I must thank Ann Middleditch for her help over the years and with these notes as well as providing a glossary. Thanks too, to John Phipps who helped with the final editing of this book.

Willie Robson, Chainbridge Honey Farm, June 2011.

Joyce Walsingham at Hexham Christmas Market 2013 with a great array of Chain Bridge Honey Farm products.

Lightning Source UK Ltd.
Milton Keynes UK
UKOW07f0417290416

273203UK00004B/4/P

9 781904 846826